THE
KID'S BOOK
OF THE
Elements

Black Dog & Leventhal Publishers
Hachette Book Group
1290 Avenue of the Americas
New York, NY 10104
www.hachettebookgroup.com
www.blackdogandleventhal.com

First Edition: October 2020

Black Dog & Leventhal Publishers is an imprint of Perseus Books, LLC, a subsidiary of Hachette Book Group, Inc.
The Black Dog & Leventhal Publishers name and logo are trademarks of Hachette Book Group, Inc.

The publisher is not responsible for websites (or their content) that are not owned by the publisher.
The Hachette Speakers Bureau provides a wide range of authors for speaking events. To find out more, go to www.HachetteSpeakersBureau.com or call (866) 376-6591.

Print book interior design by Tandem Books

Library of Congress Cataloging-in-Publication Data has been applied for.

ISBNs: 978-0-7624-7077-8 (hardcover); 978-0-7624-7078-5 (trade paperback); 978-0-7624-7079-2 (ebook)

Printed in China

1010
10 9 8 7 6 5 4 3 2 1

THE
KID'S BOOK
OF THE
Elements

AN AWESOME INTRODUCTION TO EVERY KNOWN ATOM IN THE UNIVERSE

Contents

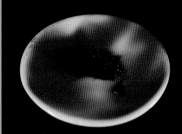

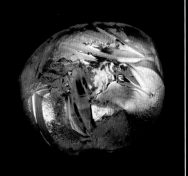

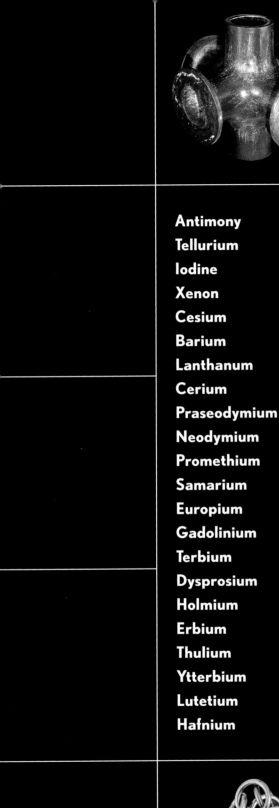

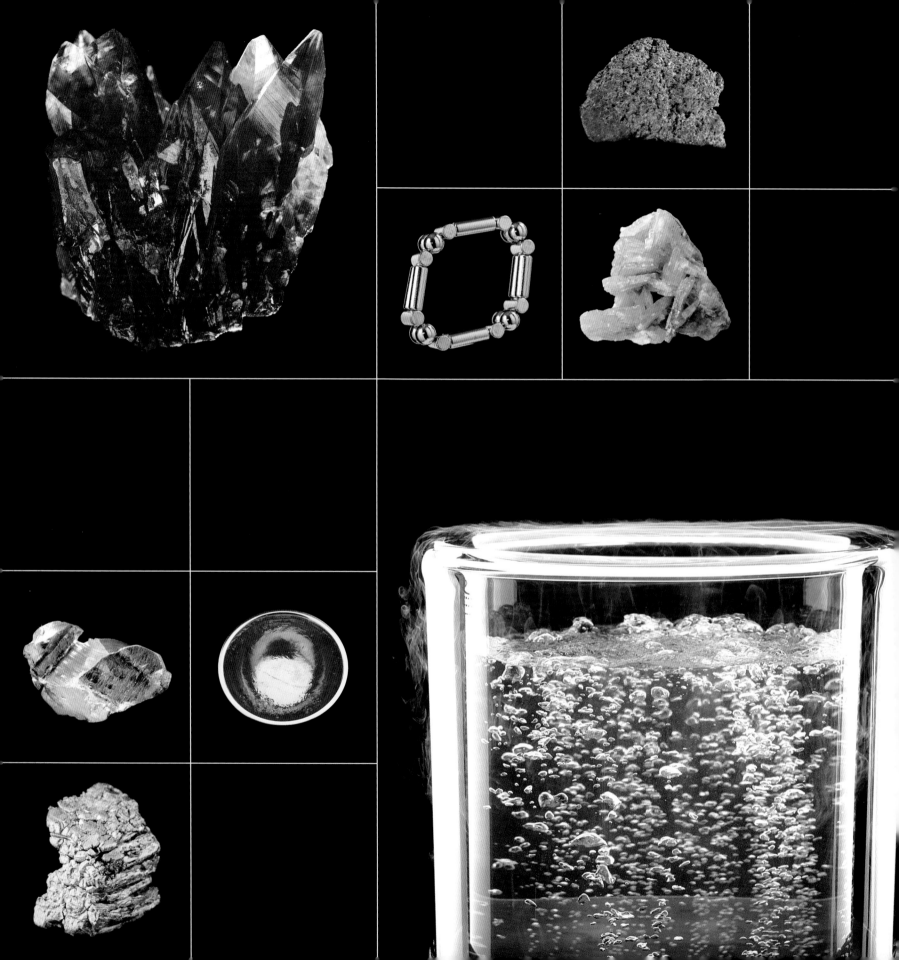

What Is the Periodic Table of the Elements?

THE PERIODIC TABLE is the universal catalog of the elements. It's a list of the building blocks that make up everything in the world that you can drop on your foot. (There are some things, such as light, love, logic, and time, that are not in the periodic table. But you can't drop any of those things on your foot.)

Earth, this book, your foot—everything in the world you can touch—is made of elements. Your foot, for example, is made mostly of oxygen with quite a bit of carbon and hydrogen joining in. Earth is made up of four main elements: iron, oxygen, silicon, and magnesium. This book is made mostly of carbon, oxygen, and hydrogen.

And this is only the beginning! In this book we'll look at all 118 elements of the periodic table.

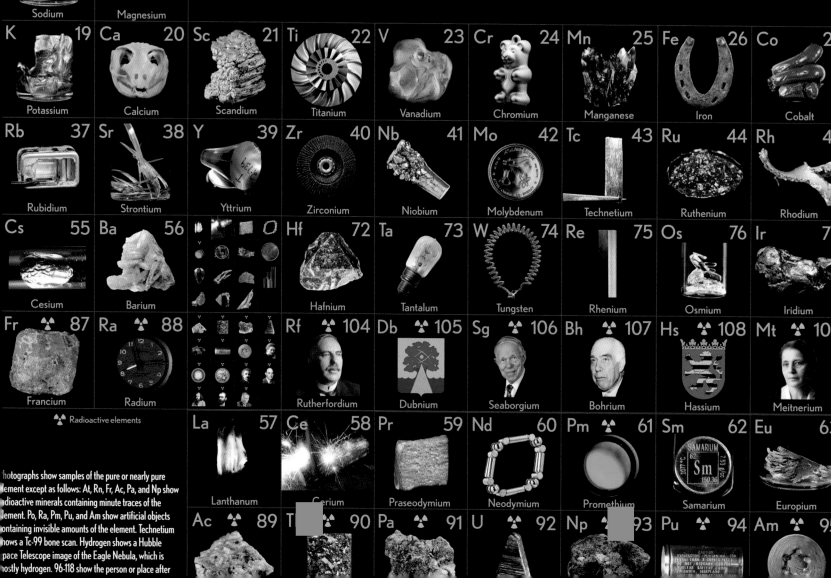

A Visual Exploration of Every Known Atom in the Universe

H Hydrogen							
Li 3 Lithium	**Be** 4 Beryllium						
Na 11 Sodium	**Mg** 12 Magnesium						
K 19 Potassium	**Ca** 20 Calcium	**Sc** 21 Scandium	**Ti** 22 Titanium	**V** 23 Vanadium	**Cr** 24 Chromium	**Mn** 25 Manganese	**Fe** 26 Iron **Co** 2 Cobalt
Rb 37 Rubidium	**Sr** 38 Strontium	**Y** 39 Yttrium	**Zr** 40 Zirconium	**Nb** 41 Niobium	**Mo** 42 Molybdenum	**Tc** 43 Technetium	**Ru** 44 Ruthenium **Rh** 4 Rhodium
Cs 55 Cesium	**Ba** 56 Barium		**Hf** 72 Hafnium	**Ta** 73 Tantalum	**W** 74 Tungsten	**Re** 75 Rhenium	**Os** 76 Osmium **Ir** 7 Iridium
Fr 87 Francium	**Ra** ☢ 88 Radium		**Rf** ☢ 104 Rutherfordium	**Db** ☢ 105 Dubnium	**Sg** ☢ 106 Seaborgium	**Bh** ☢ 107 Bohrium	**Hs** ☢ 108 Hassium **Mt** ☢ 10 Meitnerium

☢ Radioactive elements

La 57 Lanthanum	**Ce** 58 Cerium	**Pr** 59 Praseodymium	**Nd** 60 Neodymium	**Pm** ☢ 61 Promethium	**Sm** 62 Samarium	**Eu** 63 Europium
Ac ☢ 89	**T** ☢ 90	**Pa** ☢ 91	**U** ☢ 92	**Np** ☢ 93	**Pu** ☢ 94	**Am** ☢ 95

...hotographs show samples of the pure or nearly pure ...ement except as follows: At, Rn, Fr, Ac, Pa, and Np show ...dioactive minerals containing minute traces of the ...lement. Po, Ra, Pm, Pu, and Am show artificial objects ...ontaining invisible amounts of the element. Technetium ...hows a Tc-99 bone scan. Hydrogen shows a Hubble ...pace Telescope image of the Eagle Nebula, which is ...ostly hydrogen. 96-118 show the person or place after

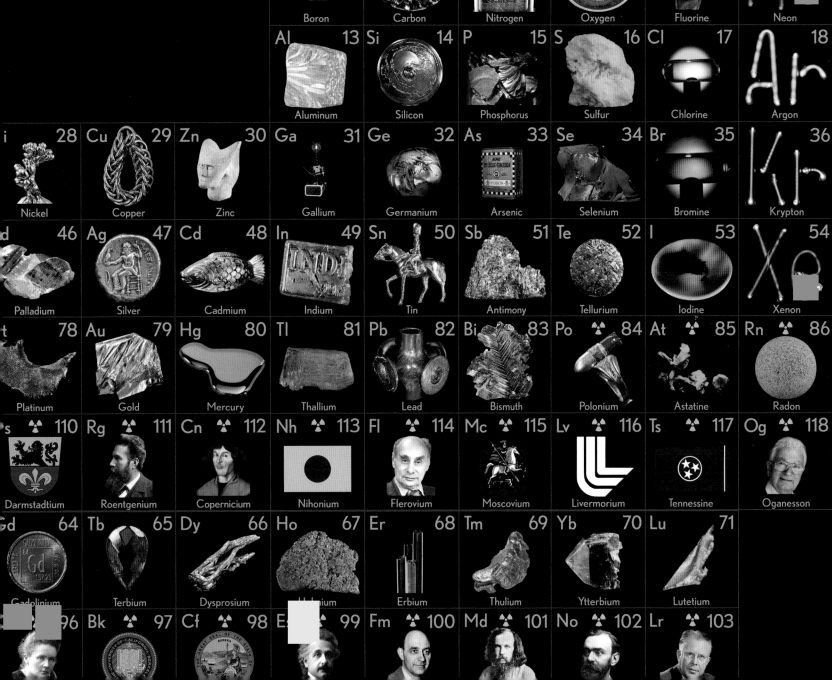

The Shape of the Periodic Table

WHY IS THE periodic table shaped the way it is? It's not a random coincidence, and it's not because it looks pretty this way. The shape of the periodic table is determined by the arrangement of electrons in the atoms of each element.

Atoms are made up of three types of "subatomic" particles, called *protons*, *neutrons*, and *electrons*. Every atom of a given element has the same number of protons in its nucleus, which is called that element's *atomic number*. For example, every oxygen atom has 8 protons in its nucleus, so the atomic number of oxygen is 8. Each atom also contains multiple neutrons, but this number isn't always the same for every atom of a given element. In orbit around the nucleus, you will find a number of electrons equal to the number of protons in the nucleus.

Reading in rows from left to right and top to bottom (sometimes with gaps in the rows), the periodic table lists the elements in order of their atomic numbers. The first row has a huge gap between hydrogen (1) and helium (2), then there are progressively smaller gaps as you go down. This might seem random, but it isn't. The gaps are there so the elements in each column share the same number of "outer shell" electrons, which are the electrons that participate in chemical bonding and thus determine the chemical properties of the elements in that column. Each column, or set of neighboring columns, forms a "group" of elements that behave in similar ways. Let's take a look at these groups.

1

3	4
11	12
19	20
37	38
55	56
87	88

THE FIRST GROUP of elements, the leftmost column, is called the *alkali metals*. The main shared property of the alkali metals is that they are fun to throw in a lake. This is because when you put an alkali metal, such as sodium (11), in water, you get a nice big explosion. I say it *can be* fun because it depends on whether or not you do it right. The result can be either thrilling or really dangerous. Chemistry is a bit like that: powerful enough to do great things in the world, but also dangerous enough to do terrible things. If you don't respect the power of chemistry, it will bite you.

Elements in the second group are called the *alkali earth metals*. They are similar to their neighbors the alkali metals, but not nearly as explosive. When you throw them into water, they react more calmly and slowly.

21 22 23 24 25 26 27 28 29 30
39 40 41 42 43 44 45 46 47 48
72 73 74 75 76 77 78 79 80
104 105 106 107 108 109 110 111 112

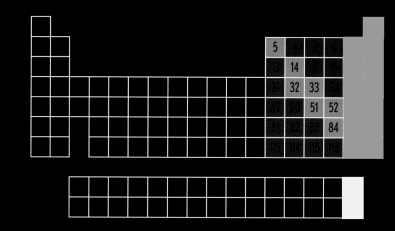

THE BIG GROUP of elements that takes up the wide middle part of the periodic table is called the *transitional metals*. When you think of metal, you probably think of something hard. All of the transitional metals, except for mercury (80), are in fact fairly hard. They are the workhorses of the periodic table—strong, stable metals used to make everything from airplanes to skyscrapers.

Notice the two empty spaces in the lower-left corner of this group? Those spaces are reserved for the *lanthanide* and *actinide* groups of elements, which we will talk about in a minute.

THIS NEXT BLOCK of elements represents three different groups.

The elements in the lower-left red triangle are known as the *ordinary metals*. The group of rust-colored elements in the upper-right triangle is known as the *nonmetals*. The orange group of elements in between is known as the *metalloids*, because these elements are kind of like metals and kind of not like metals.

THE SECOND-TO-LAST COLUMN of elements in the periodic table is called the *halogens*. This is a particularly nasty group, known for being violent and smelly when they're out in the world on their own. But when they mix with elements from other groups, they can transform into pretty tame things—toothpaste and table salt, for example.

The elements that make up the last column are known as the *noble gases*. Sounds fancy, eh? The name comes from the fact that, like kings and other royalty, they don't mix with the commoners from the rest of the table. Except for occasionally hanging out with fluorine (9), they form no chemical compounds.

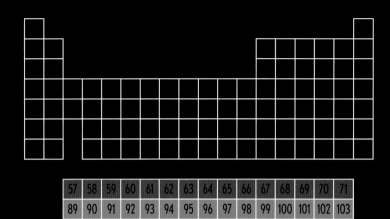

THESE TWO GROUPS of elements sit just below the main periodic table and are known collectively as the *rare earths*. The top row is the *lanthanides* and the bottom row is the *actinides*. (These are the elements that fill up those two spaces left in the block of the transitional metals.)

Sometimes you will see the periodic table arranged with the rare earths slotted into those blank spaces, but then the table becomes very wide. So usually you see the rare earths arranged as two separate rows at the bottom.

NOW YOU'VE SEEN the periodic table as a whole, and you know how it's arranged. In the rest of this book you will get to meet each of the 118 elements as individuals.

Glossary

ALLOY: A mixture of different metal elements.

ANODIZE: To send an electric current through the surface of a metal to form a hard, protective oxide coating.

ATOM: A tiny particle of matter made up of a nucleus (containing protons and neutrons) with electrons in orbit around it. Each atom is an atom of a particular element, determined by the number of protons in its nucleus. For example, an atom with six protons is a carbon atom; this is called the *atomic number* of the element.

COMPOUND: A substance containing two or more different chemical elements that are chemically bonded to each other in specific ratios (unlike an alloy, where there is no fixed proportion in the number of atoms of each type).

ELECTRON: A tiny, subatomic (smaller than an atom) particle with a negative electric charge. Electrons are responsible for creating the chemical bonds between atoms that form compounds.

EMISSION LINE: Light of a specific wavelength (color) emitted by an atom or compound when it is heated in a flame or electric arc.

HALF-LIFE: The length of time before half the atoms in a sample of a radioactive element will undergo radioactive decay (meaning energy from the nuclei gets released).

INERT: Resistant to involvement in chemical reactions.

ISOTOPE: Different isotopes of the same element are atoms of that element that have the same number of protons in their nucleus, but different numbers of neutrons.

NEUTRON: A subatomic (smaller than an atom) particle with zero electric charge and found in the nucleus of most elements.

ORE: A raw material, dug from the ground, that can be transformed into a metal by chemical or electrical means. Most ores are oxides: for example, iron ore is iron oxide, and aluminum ore is aluminum oxide.

OXIDE: A compound of oxygen and another element.

PIGMENT: A chemical compound with a particularly intense color that can be used in paints or inks.

PROTON: A subatomic (smaller than an atom) particle with a positive electric charge. The number of protons in the nucleus of an atom determines what element that atom is. This is called the *atomic number* of the element.

REACTIVITY: How easily an element or compound will engage in chemical reactions. Highly reactive chemicals can be dangerous, because they tend to react rapidly and release a lot of energy when they encounter other chemicals, including things like your skin or lungs.

SPECTRAL LINE: See *Emission Line.*

SUPERCONDUCTOR: A material that allows electricity to flow through it without any resistance.

▷ By weight, 75% of the visible universe is hydrogen. Usually, it is a colorless gas, but large quantities of it in space actually absorb starlight, creating spectacular sights such as the Eagle Nebula, seen here in an image captured by the Hubble Space Telescope.

Elemental

Hydrogen

HYDROGEN IS THE lightest of all the gases—even lighter than helium (2), which makes balloons float. It is also the most abundant element in the universe. Our sun consumes 600 million tons of hydrogen per second, converting it into 596 million tons of helium. Yes, *600 million tons per second*. Even at *night*! The leftover 4 million tons per second is converted into energy by Einstein's famous formula $E = mc^2$; it becomes the light and heat of the sun that sustains us all.

He

2

Elemental

Helium

HELIUM IS NAMED for the Greek god of the sun, Helios, because the first clues of its existence were dark lines in the spectrum of sunlight that could not be explained by the presence of other known elements. This made helium the first element to be discovered outside Earth. It has since been found down here, where we put it in party balloons.

Li

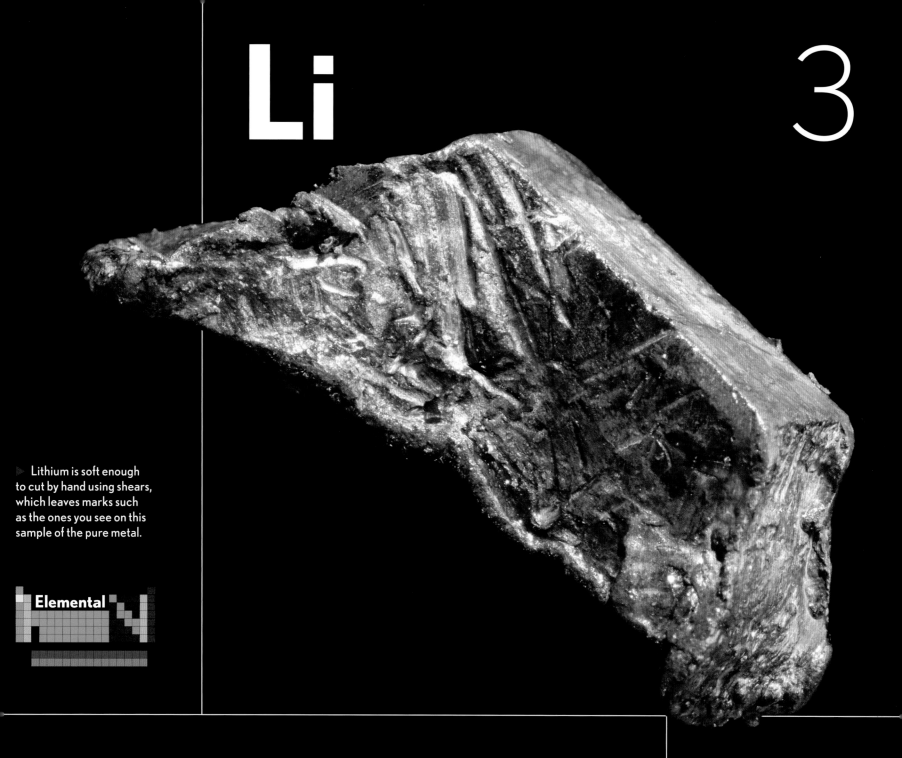

▷ Lithium is soft enough to cut by hand using shears, which leaves marks such as the ones you see on this sample of the pure metal.

Elemental

LITHIUM IS A very soft, very light metal. So light it actually floats on water! Lithium in lithium-ion batteries powers all kinds of electronic devices, from cell phones to electric cars.

Lithium

Be

◁ This pure, broken crystal of refined beryllium would ordinarily be melted down and turned into strong, lightweight parts for missiles and spacecraft.

Beryllium

BERYLLIUM IS A very light, very strong metal that's resistant to corrosion. It's also very expensive. Beryllium is used to make parts for missiles and rockets, where no one seems to care about the cost, and where strength without weight is the most important thing to think about.

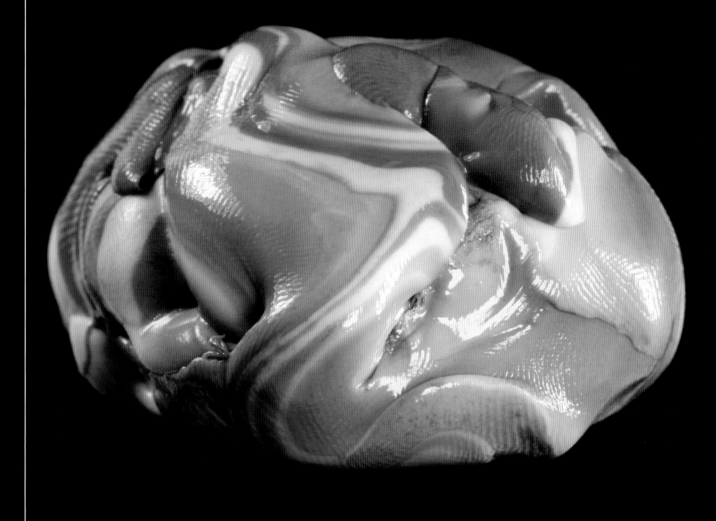

▷ Silly Putty gets its bounce from boron cross-links in its chemical structure.

Elemental

WITH A NAME like "boron," how can it get any respect? Even though it has a funny name, boron has lots of cool uses. I promise! For example, it's the critical element that gives Silly Putty its amazing ability to be soft and moldable in your hand, yet hard and bouncy when you throw it against the wall. A particular form of boron nitride might one day prove to be harder than diamond, which is one of the hardest substances found on Earth.

Boron

C

6

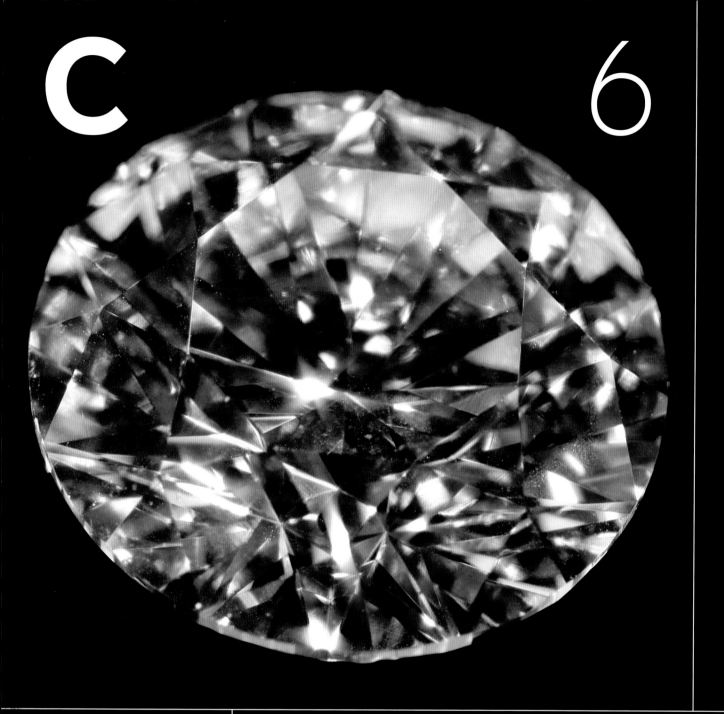

Carbon

CARBON IS *THE* most important element for life, period. It forms the spiral backbone of DNA and is therefore the backbone of all life on Earth. It also forms diamonds. Almost every molecule in your body, except water and your bones, has carbon atoms in it: fat, protein, enzymes, oils, hair, fingernails, poop, the works. They're all made of organic compounds, all of which have carbon as their main structural element.

▷ A vacuum flask filled with liquid nitrogen that is boiling at –320° Fahrenheit (–196° Celsius).

Elemental

OVER 78% OF Earth's atmosphere is nitrogen. The other 22% is mostly the oxygen (8) we need to breathe. Nitrogen is found in many compounds, but in the air it's in the form of N_2, which is quite difficult to break apart. One of the most important advances in chemistry was the discovery of a way to take nitrogen out of the air and turn it into nitrogen-rich fertilizer. A good fraction of Earth's population would not exist without the food grown with this nitrogen.

Nitrogen

O

8

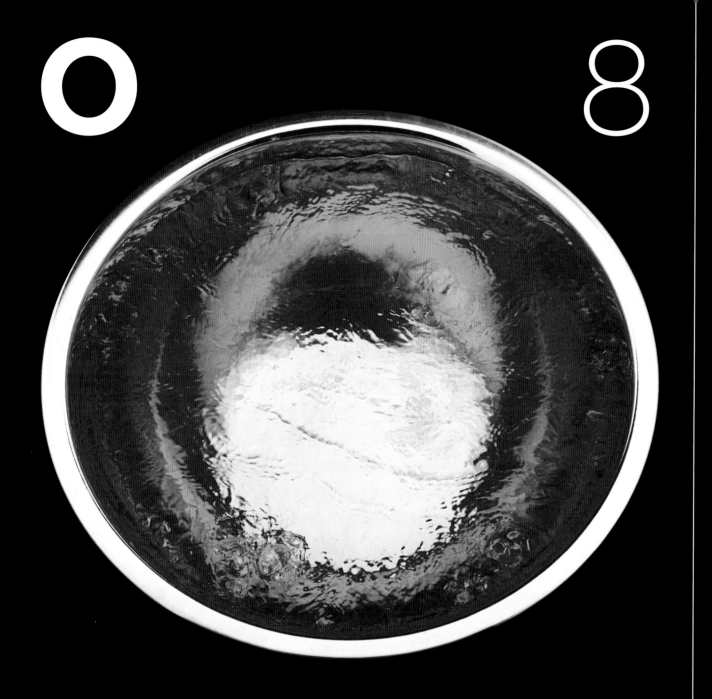

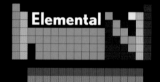

Oxygen

OXYGEN IS THE most abundant element on Earth. It's nearly half of the weight of Earth's crust and 86% of the oceans. We also need it to survive. You can live a month without food and a week without water. But a few minutes without oxygen, forget it. As we explore space, I'm sure that oxygen will become almost mythical. It's the one thing you absolutely must have, and if it's running low on your ship, nothing else matters.

F

9

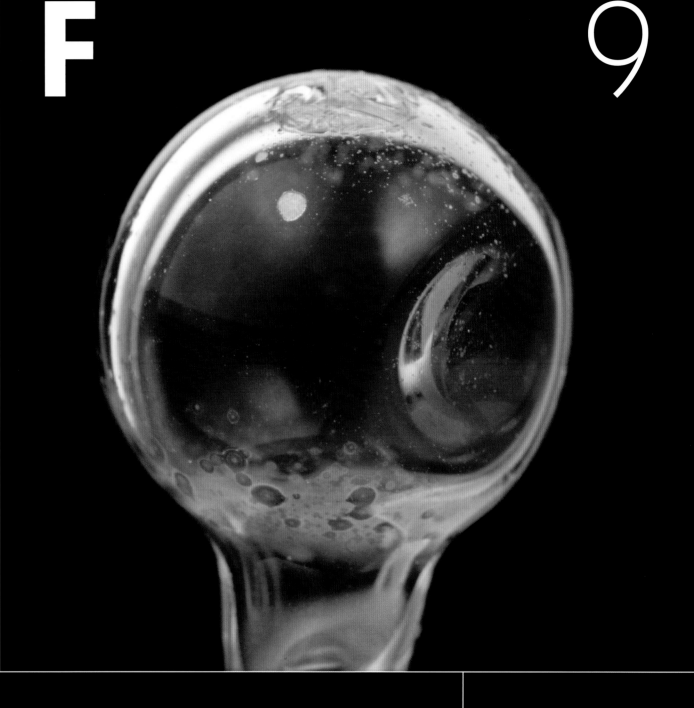

Elemental

FLUORINE IS THE most reactive of all the elements. If you blow a stream of fluorine gas at almost anything, it will burst into flames. But the fierce way it reacts means that compounds formed with fluorine are very chemically stable.

For example, Teflon (used on nonstick pans) is full of carbon-fluorine bonds. A lot of energy was released when those bonds formed, so it takes a lot of energy to break them. Teflon is almost impossible to attack chemically.

Fluorine

Ne

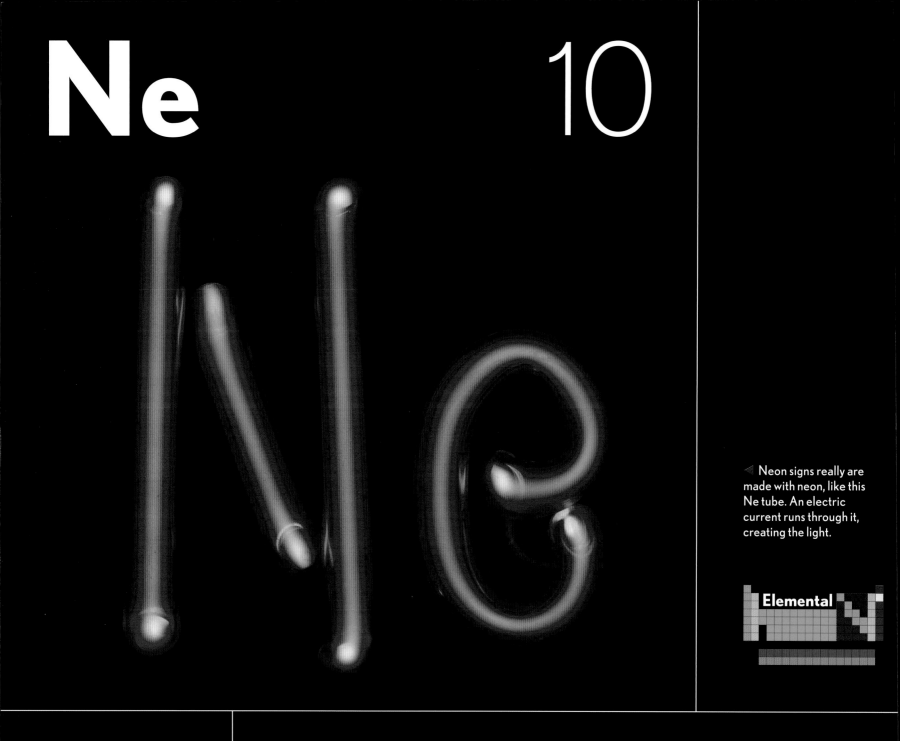

◀ Neon signs really are made with neon, like this Ne tube. An electric current runs through it, creating the light.

Elemental

Neon

NEON IS THE least reactive of all the elements. That means it completely refuses to form compounds with any of the other elements. You may not be able to do anything chemical with neon, but it is very good at making bright red-orange light when you run electric current through it. Although LEDs are taking over the world, in the past, neon light gave places like Times Square in New York and Las Vegas, Nevada, their flashy character.

Na

▷ These soft, silvery sodium chunks were cut with a knife and stored under oil. In the open air, they would turn white in seconds. If put in water, they would generate hydrogen (1) gas and explode in flaming balls of molten sodium.

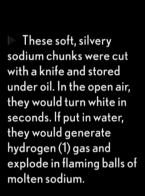
Elemental

SODIUM IS THE best of the alkali metals for making explosions. It's also the best tasting, because when you combine it with chlorine (17) you get table salt!

Sodium

Mg

An engraved magnesium printing block.

Elemental

Magnesium

MAGNESIUM IS A good metal for making lightweight bicycle parts, but grind it into a fine powder and it's highly flammable. Old-fashioned camera flashes used magnesium, and many modern fireworks use magnesium powder to create a bright white light.

Al

13

You can see the internal crystal structure in this etched, high-purity aluminum bar.

Elemental

ALUMINUM DOESN'T RUST when it comes in contact with air, which gives it a big advantage over iron (26) despite being more expensive. Even though it doesn't rust, it does react with air. The product of that reaction is a super-hard, transparent coating that protects the metal from any further decay. Problem solved!

Aluminum

Si

14

This silicon boule (a synthetic, human-made mass in the form of a single, nearly perfect crystal) was pulled out of the melting pot before it was done growing. We're seeing the underside, where molten silicon dripped off.

Silicon

COMPUTER CHIPS THAT power phones, laptops, Instasnap, Facechat, and so on—basically life as we now know it—start out as common white silica beach sand, otherwise known as *silicon dioxide*. The bones of the earth—the

rock, sand, clay, and soil—are made up in large part of silicate minerals. So if silicon-based artificial intelligences ever take over, they will have no shortage of raw material to replicate themselves.

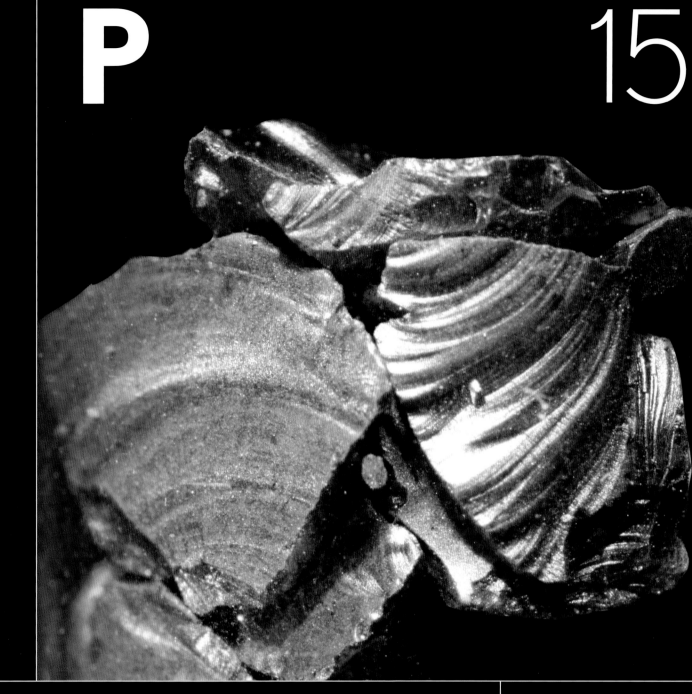

▷ This rare violet phosphorus is a mixture of red and black phosphorus, which means it's not a true allotrope (the name for different forms of the same element).

Elemental

PHOSPHORUS EXISTS IN various forms called *allotropes*. The tips of matchsticks are often made with red phosphorus. Black phosphorus is hard to make and rarely seen since it has no important uses. White phosphorus is a deadly poison and burns on contact with air. It is used mainly in warfare as a weapon of pure evil.

Phosphorus

S

◁ Rocks of pure sulfur occur naturally around volcanoes and geothermal vents, where the internal heat of the Earth breaks down sulfur-bearing minerals and releases the pure element.

Elemental

Sulfur

SULFUR IS SMELLY! You probably know it best from its compound hydrogen sulfide, which smells like rotten eggs. But this smelly element is also very useful. Vast quantities of it are produced and consumed in the chemical industry, mostly in the form of sulfuric acid, the workhorse acid used in countless manufacturing processes.

Cl

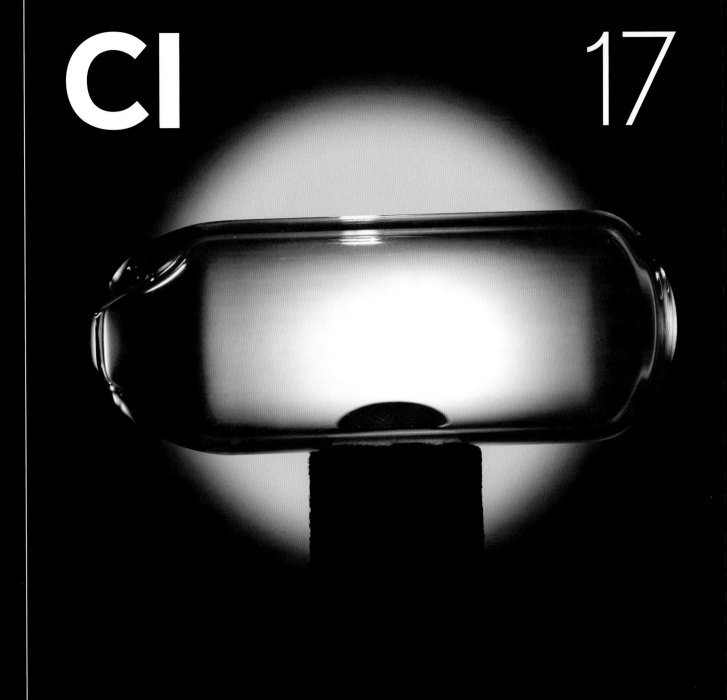

▶ Chlorine gas has a pale-yellow color that you can just see against the white background.

CHLORINE WAS USED in large quantities as a deadly gas during World War I, but in small amounts it is one of the cheapest, most effective, and least harmful disinfectants, saving millions of lives by cleaning up drinking and wastewater with no long-lasting environmental effects. If you think about it that way, chlorine has saved many more lives than it has taken.

Chlorine

Ar

18

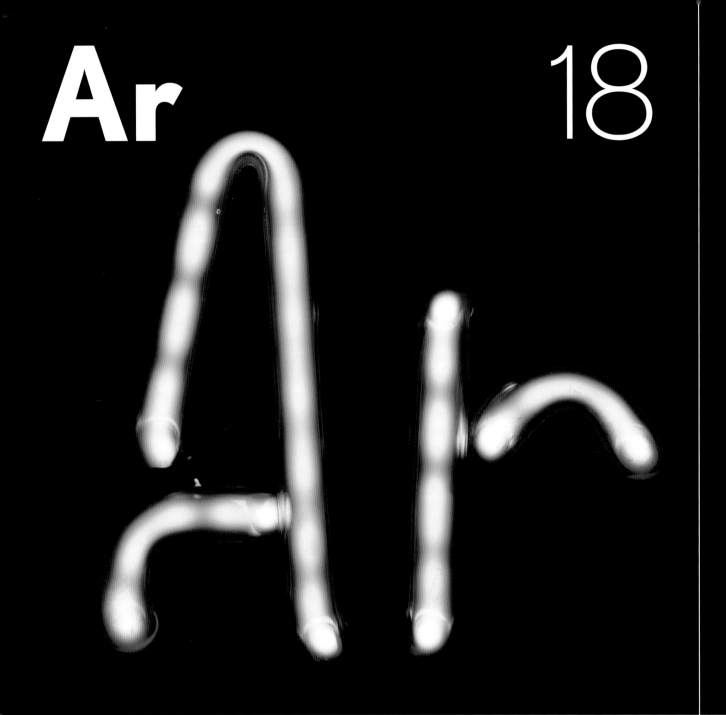

As a noble gas, argon is inert and colorless until an electric current excites it, giving it a rich, sky-blue glow.

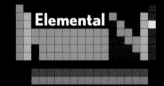

Elemental

Argon

ARGON IS SURPRISINGLY abundant in Earth's atmosphere, which is why it's also relatively cheap. One of its most common uses was in old-school incandescent lightbulbs, which were filled with a mixture of nitrogen (7) and argon. It's still the noble gas of choice for shielding chemical reactions that would be messed up by contact with air. For this, people work in argon-filled boxes by reaching in with rubber gloves mounted below a window in the box.

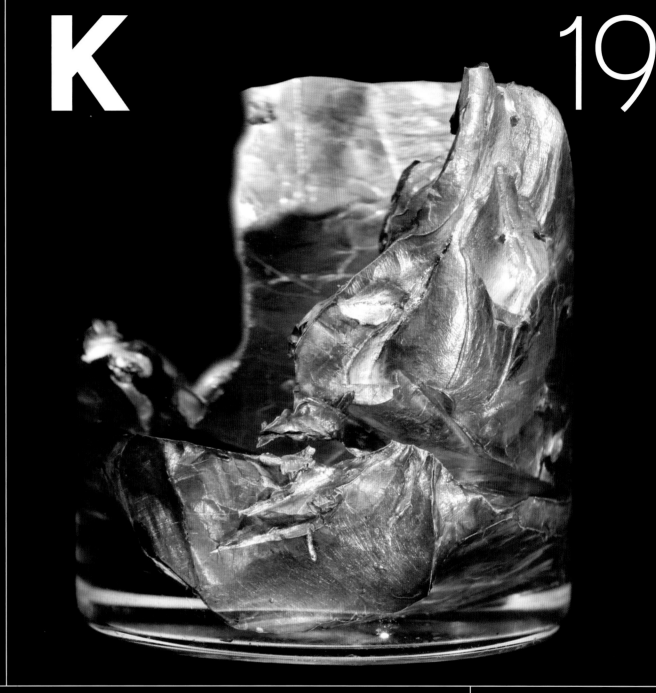

▷ The purple tint on these soft potassium cubes is a very thin oxide coating. They turn black in seconds when exposed to air. When exposed to water, they explode.

Elemental

BANANAS ARE HEALTHY and a rich source of the important nutrient potassium, without which we cannot survive. About 1 in every 10,000 potassium atoms in the world is the radioactive isotope ^{40}K. Thus, bananas are both healthy *and* radioactive.

Potassium

Ca

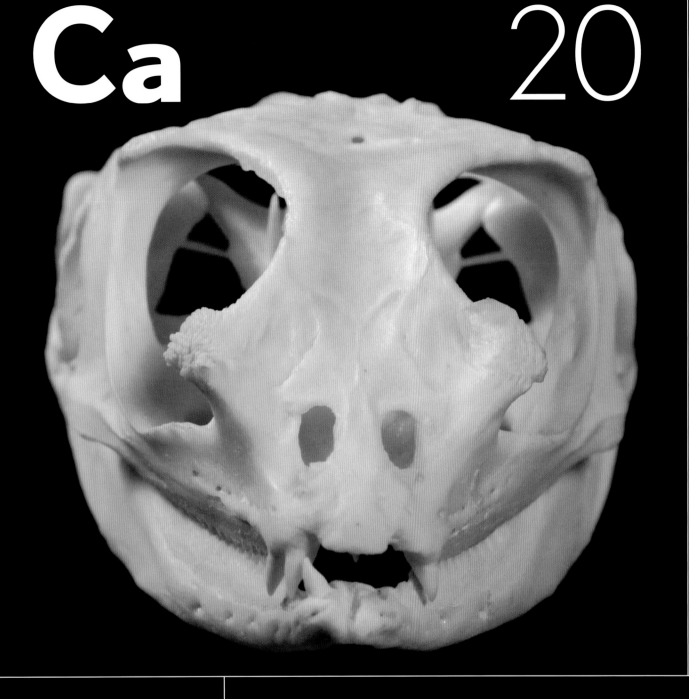

This frilled dragon skull is made of hydrated calcium phosphate, the same material that makes up your bones.

Elemental

Calcium

WHEN YOU THINK of calcium, you probably think of white chalky things or perhaps milk. Both chalk and milk have calcium compounds (combinations of calcium and another element). Pure

calcium is actually a shiny metal that looks kind of like aluminum (13). But you rarely see the pure element because it has few uses in pure form, and it quickly gets a black coating when exposed to air.

Sc

21

These scandium crystals were grown to eventually be turned into scandium salts to be used in daylight-spectrum metal halide arc lights.

Elemental

SCANDIUM IS USED to make strong metals. Just a tiny amount mixed with aluminum (13) creates some of the strongest aluminum alloys known. These are used to build fighter jets, baseball bats, and bicycle frames.

Scandium

Ti

22

◀ This titanium "blisk," short for *bladed impeller disk*, is from a small jet engine.

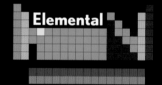

Titanium

TITANIUM IS NAMED after the Titans, the gods of Greek legend who stand for strength. It's used to make jet engines and power tools. It's also completely nonrusting and nonallergenic, so it can be used to make everything from artificial hip joints and dental implants to earrings and other jewelry.

V

23

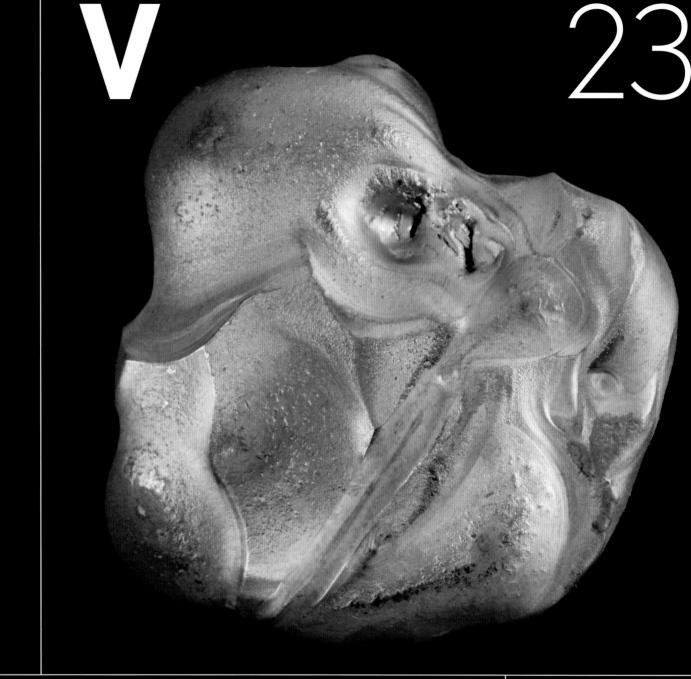

Vanadium

▷ Pure, molten vanadium has a lovely surface.

Elemental

VANADIUM IS ONE of the hardest, toughest, and most wear-resistant of all the elements. It's heavier than titanium (22) and much harder. It's mainly used in alloys like chrome-vanadium steel to make tools. Look at your average crescent wrench and you'll probably see it stamped Cr-V for those two elements. The tool is still at least 90% iron (26), but it's the alloying elements that give different kinds of steel the unique properties that make them work well for particular uses.

Vanadium

Cr 24

Chromium

A MICROSCOPICALLY THIN layer of chrome plating is all you normally see of this element in its pure form, but when mixed with iron (26) and nickel (28), chromium is important for making stainless steel. It's extremely shiny, highly resistant to corrosion, and beautiful in so many ways.

Mn

25

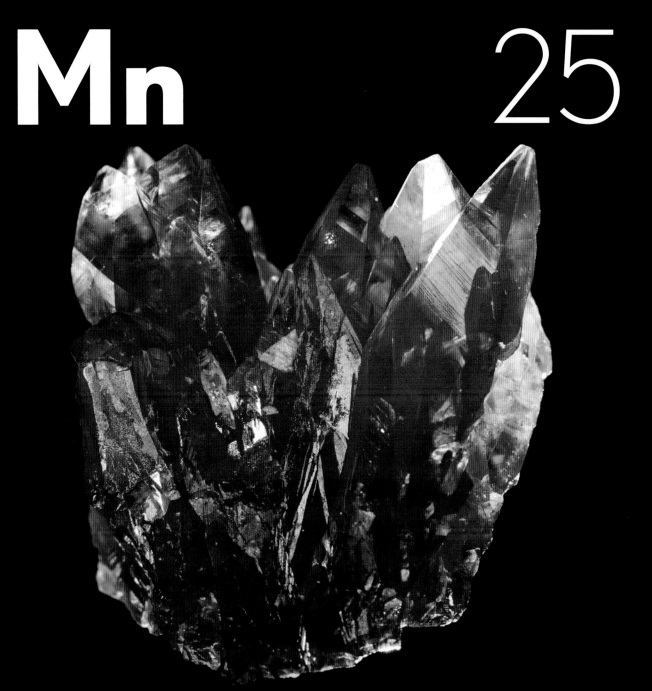

▷ I traded a mineral dealer this magnificent rhodochrosite (manganese carbonate) crystal for several hundred lesser minerals.

Elemental

ALONG WITH RED iron oxide, black manganese oxide was one of the first pigments ever discovered. It's been found in cave paintings that are at least 17,000 years old. The pure metal tarnishes too quickly to be useful today, but when it's mixed with iron (26), it makes a steel alloy with particularly strong "work hardening" properties. A safe or vault door made of manganese steel is difficult to drill into because the more you drill, the harder the steel becomes.

Manganese

Fe

26

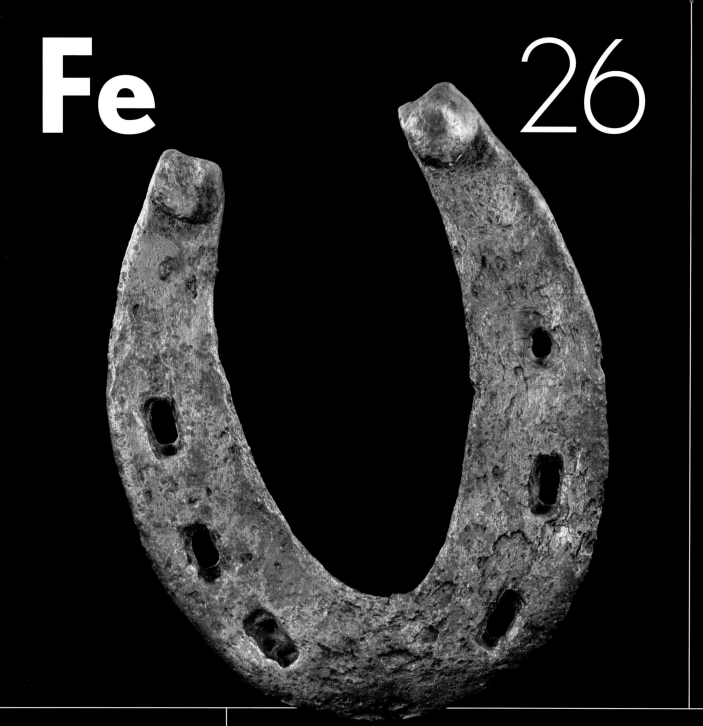

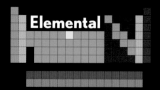

◄ This medieval horseshoe has pits (the small red, pitted areas you see in the metal) from centuries of slow rusting.

Elemental

Iron

IRON IS THE only element to have an age named after it, as in the "Iron Age." It richly deserves the honor for being the primary toolmaking material then and now. Today, iron is very cheap and forms a huge range of alloys. It can be welded, cast, machined, forged, cold-worked, tempered, hardened, annealed, drawn, and generally persuaded to take on all kinds of shapes. The only downside to iron is that it rusts so easily, which is just one of the great lousy breaks of chemistry.

Co

27

Cobalt-aluminum oxide has been an important pigment in paints for centuries.

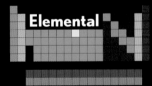

Cobalt

IN ITS PURE form, cobalt is a quite ordinary gray metal that looks sort of like nickel (28). But cobalt-aluminum oxide is a stunning blue pigment that has been used for centuries by artists and artisans to make ceramics, jewelry, and paint.

Ni

28

◁ Nickel-chrome nodules like this grow where the insulation on electroplating racks has cracked. They are a beautiful nuisance to the plating industry.

Elemental

Nickel

NICKEL IS WIDELY used to make coins, which is why it makes sense that there is a U.S. coin called a *nickel*. But

U.S. nickels are actually only about 25% nickel; the rest is copper (29).

Cu

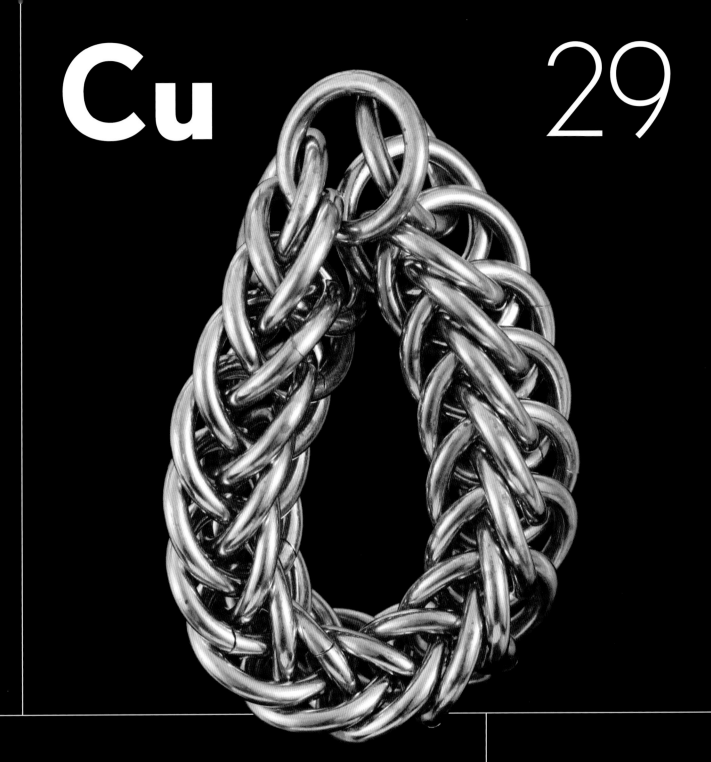

▷ A woven chain made from copper electrical wire.

Elemental

COPPER IS SOFT enough to be molded using hand tools, yet it's hard enough to be made into very useful things—from pots and pans to jewelry, pipes, and other hardware. Copper is especially useful when it is alloyed with tin (50) to create bronze or zinc (30) to create brass.

Copper

Zn

I made this zinc casting when I was a kid.

Elemental

Zinc

ZINC IS A cheap and easy metal to cast, especially for things that don't need to be particularly strong. U.S. pennies are now made primarily of zinc. They used to be made of copper (29), until the price of copper became greater than the value of the penny itself.

Ga

31

Gallium

▷ A Blu-ray gallium nitride laser diode in operation.

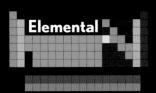

Elemental

LIKE MERCURY (80) and cesium (55), gallium has an extremely low melting point. Even in Alaska, a small chunk of gallium will literally melt in your hand. But this is an experience you're not likely to repeat, because it will stain your skin a dark brown color. It's best to keep your gallium in a plastic bag when you play with it.

Gallium

Ge

Molten germanium forms crystals on its surface as it cools.

Elemental

Germanium

GERMANIUM WAS DISCOVERED nearly 20 years after Dmitri Mendeleev predicted its existence and established the shape of the periodic table in 1869. It was discovered by Clemens Winkler and bears the name of his homeland, Germany. Although most computer chips are made of silicon (14), the very fastest ones are made with germanium.

As

▶ Paris green, which is made from arsenic, is equally useful in wallpaper pigments and rat poison.

Elemental

ARSENIC, IN THE form of a compound called *Paris green*, is an insecticide and rat poison. But in 19th-century England, it was also used to dye fancy wallpaper. Unfortunately, in damp winters mold grew on the wallpaper, turning the ink into a poisonous gas. People started to think that getting outside more or moving to a dry climate was healthy. No surprise, really. If you're being poisoned by your wallpaper, getting out of the room is probably a good idea no matter where else you go!

Arsenic

Se

34

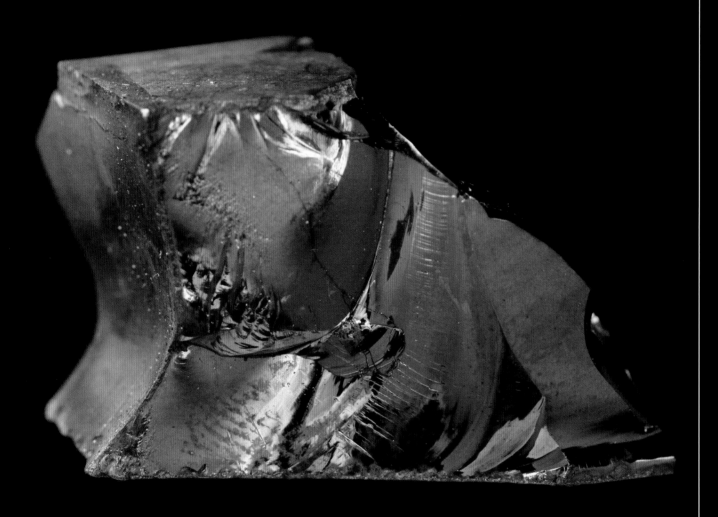

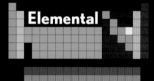

Elemental

Selenium

SELENIUM CAN BE made to respond to light by changing it from blocking the flow of electricity to allowing it to flow. Laser printers have a cylinder coated with this form of selenium. It's first charged with static electricity, then exposed to an image.

Where the image is light, the selenium becomes conductive, and the static electricity is neutralized. Fine black powder sticks only where there is still static charge, then heat and pressure transfer the powder onto paper pressed against the drum.

▶ Bromine is liquid at room temperature, but it evaporates very rapidly into a deep reddish-purple gas.

Elemental

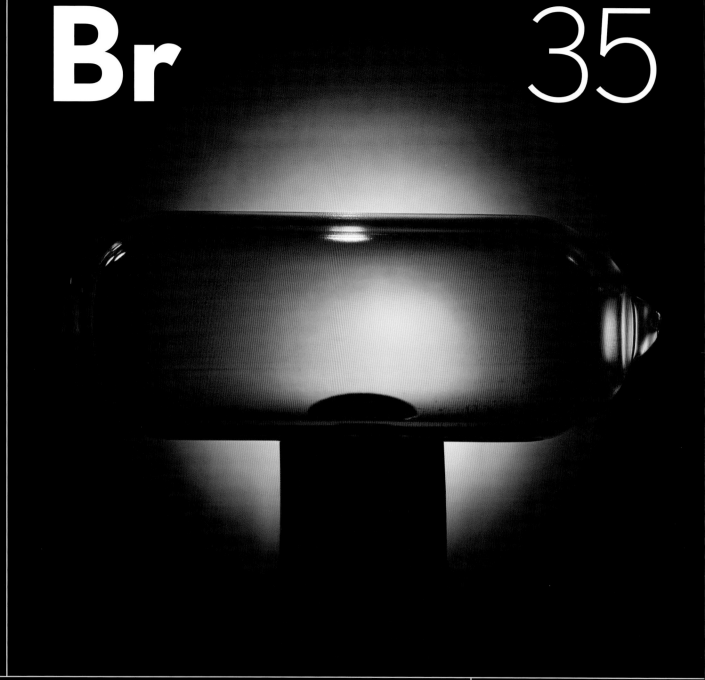

EXACTLY TWO STABLE elements are liquid at room temperature, mercury (80) and bromine. But bromine's boiling point is so low that even at room temperature a puddle of it will evaporate away in a cloud of reddish-purple vapor in less than a minute. Like chlorine (17), which is just above it in the periodic table, bromine vapor smells violently, painfully bad and will kill you by dissolving your lungs from the inside if you don't get away from it fast enough.

Bromine

Kr

36

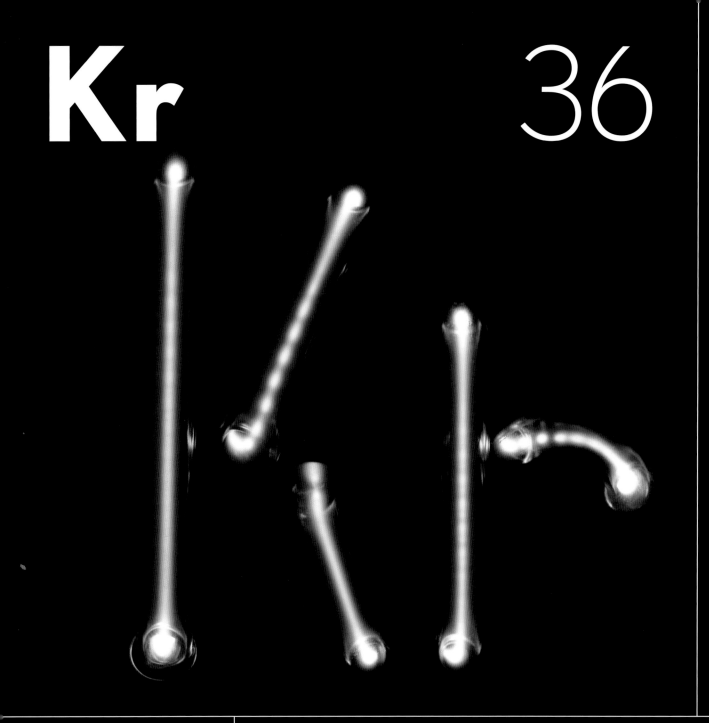

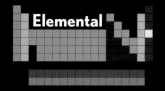

Krypton

LIKE ALL THE other noble gases, krypton refuses to bond with any other element. This makes it handy when you want to protect something from the rest of the world. Higher-efficiency incandescent bulbs are filled with krypton gas, which reduces evaporation of the tungsten (74) filament inside, allowing the bulb to operate longer at higher temperatures. (Nobody really cares about this anymore because LEDs are a much better alternative.)

A complete rubidium clock cell, less than an inch wide, includes a rubidium vapor cell, heating coils, and transmitter and receiver antennas.

Elemental

RUBIDIUM IS NOT related to rubies, but both get their name from the Latin for "red." While rubidium itself is not red, it was first discovered as an unexplained red line in an emission spectrum. (An emission spectrum is the range of light released by an element when it is made extremely hot. Each element releases its own unique spectrum.) There are precious few uses for rubidium, but it's that spectral line that accounts for the purple color of some fireworks.

Rubidium

Sr

38

This pure strontium metal is slightly oxidized even though it's stored under mineral oil.

Strontium

THE RADIOACTIVE STRONTIUM isotope ^{90}Sr, associated with the fallout from nuclear explosions, is the black sheep of the strontium family and has unfairly tainted this element's reputation. Ordinary strontium is not radioactive at all. In fact, the active ingredient in some toothpaste is strontium acetate.

Y

▷ A YAG (yttrium aluminum garnet) laser crystal boule.

Elemental

Yttrium

THE COMPOUND YTTRIUM barium copper oxide (commonly known as *YBCO*) turns into an amazing, almost magical superconductor when cooled with liquid nitrogen (7). If you try to sit a magnet on top of a cooled YBCO disk, you can't do it, because the magnet will stop about a quarter of an inch above the disk and just float in midair, plain as day. It's freaky!

Zr

40

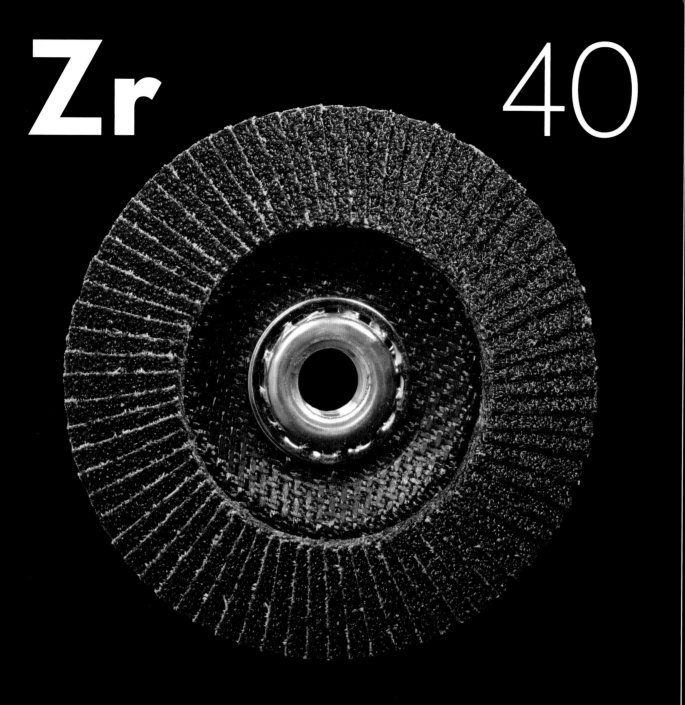

Cubic zirconia, the crystal form of zirconium oxide, is used as fake diamond and as an important industrial abrasive. You can see it on this flap wheel used by welders.

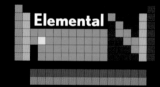

Zirconium

ZIRCONIUM IS A tough, hard metal. In the form of the compound zirconium oxide, it's used for grinding wheels on oil rigs, giant earth-moving equipment, and dirt bikes. But it has a softer side, too. In cubic crystal form, it's known as *cubic zirconia* (CZ), which is by far the most common type of fake diamond used for engagement rings and other jewelry.

Nb

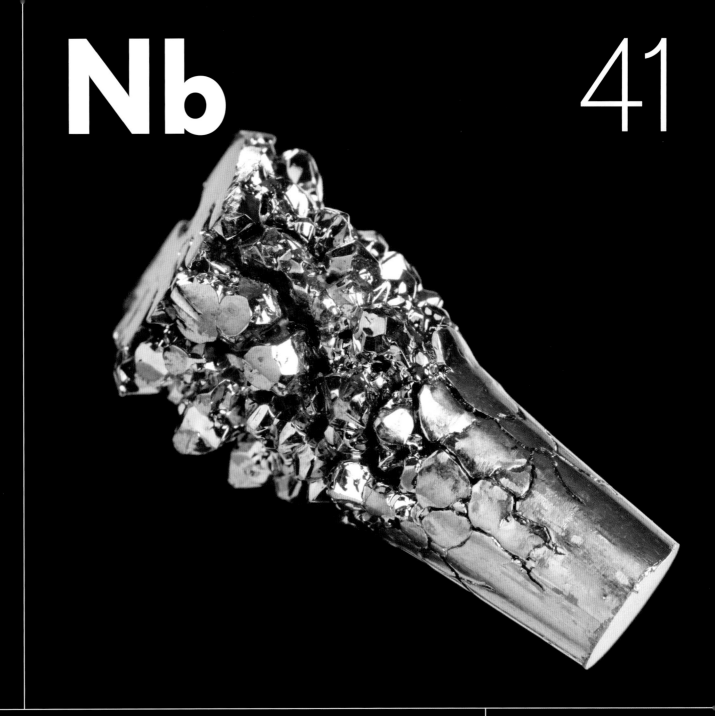

A high-purity crystal niobium bar.

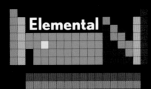

NIOBIUM IS NAMED after Niobe, the granddaughter of Zeus in Greek mythology. Rocket nozzles are made of niobium alloys because they resist corrosion even at high temperatures.

Jewelry and coins are also made of niobium because it can be beautifully anodized, which forms a rainbow of colors on its surface.

Niobium

Mo

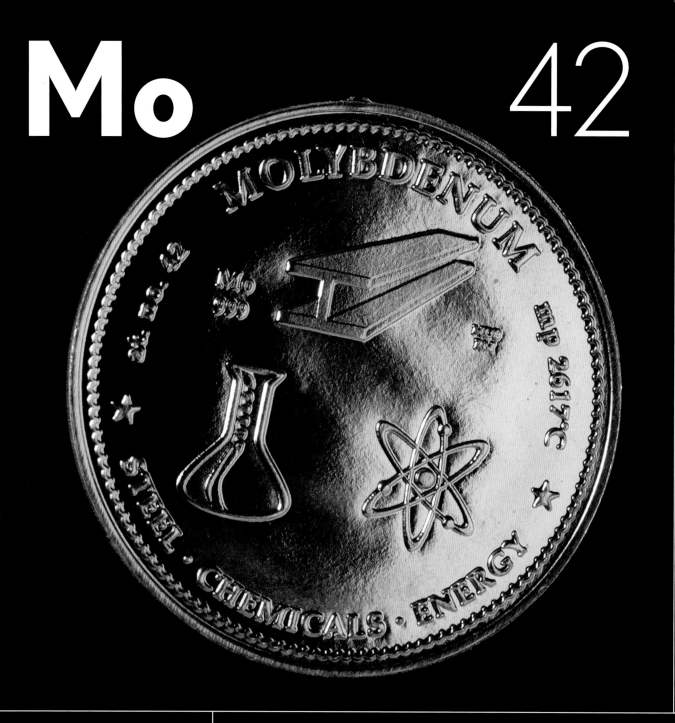

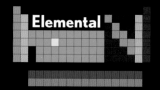

Molybdenum is not often used to make coins. This medallion was made in honor of a molybdenum mine.

Elemental

Molybdenum

MOLYBDENUM IS A metal of industry through and through. It is often used to add great strength and heat resistance to steel alloys, most notably to M-series high-speed tool steels (*M* is for *molybdenum*, of course).

Tc

43

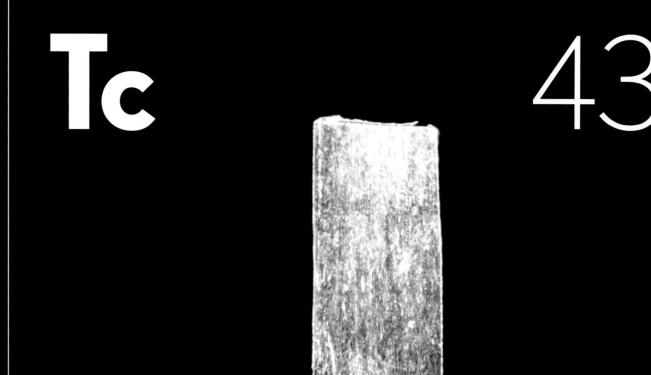

Technetium

TECHNETIUM IS A radioactive
element located, surprisingly, smack
in the middle of the most stable
neighborhood of the periodic table.
It got its name because it was the first

non-naturally occurring element to be
created. It exists only through technology
(except for vanishingly small amounts in
certain ores).

Ru

Ruthenium chloride is a vivid red.

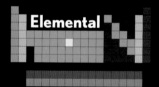

Elemental

Ruthenium

RUTHENIUM IS A precious metal, similar to platinum (78). In daily life, you're most likely to find ruthenium in jewelry as a thin plating that has a darkish gray, pewter-like shine to it.

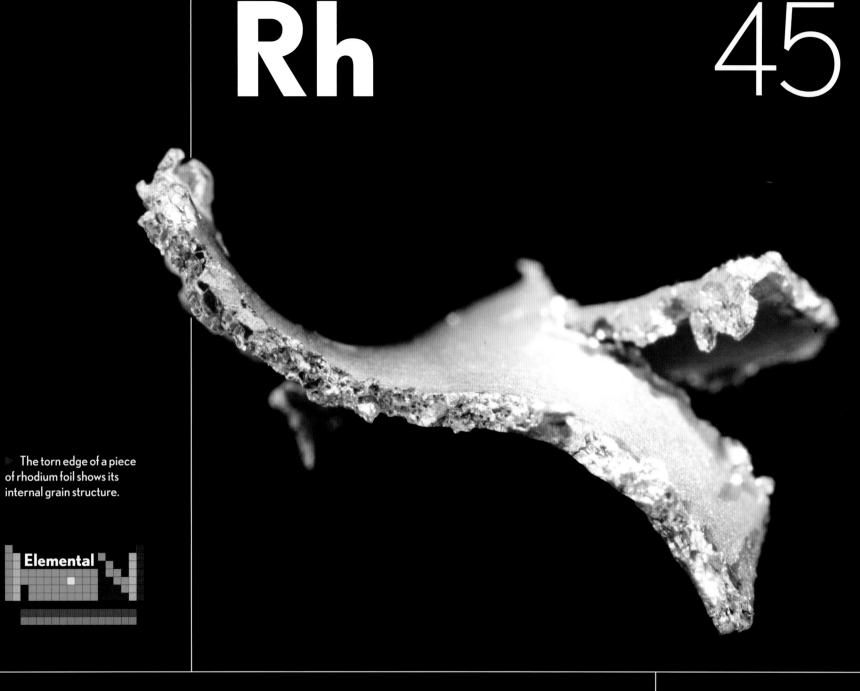

Rh

45

The torn edge of a piece of rhodium foil shows its internal grain structure.

Elemental

Rhodium

RHODIUM IS FAMOUS for being shiny. Costume jewelry designed to look like silver (47) or platinum (78) is often plated in rhodium because a film of rhodium just one micron thick is shinier than all the platinum in the world.

Pd

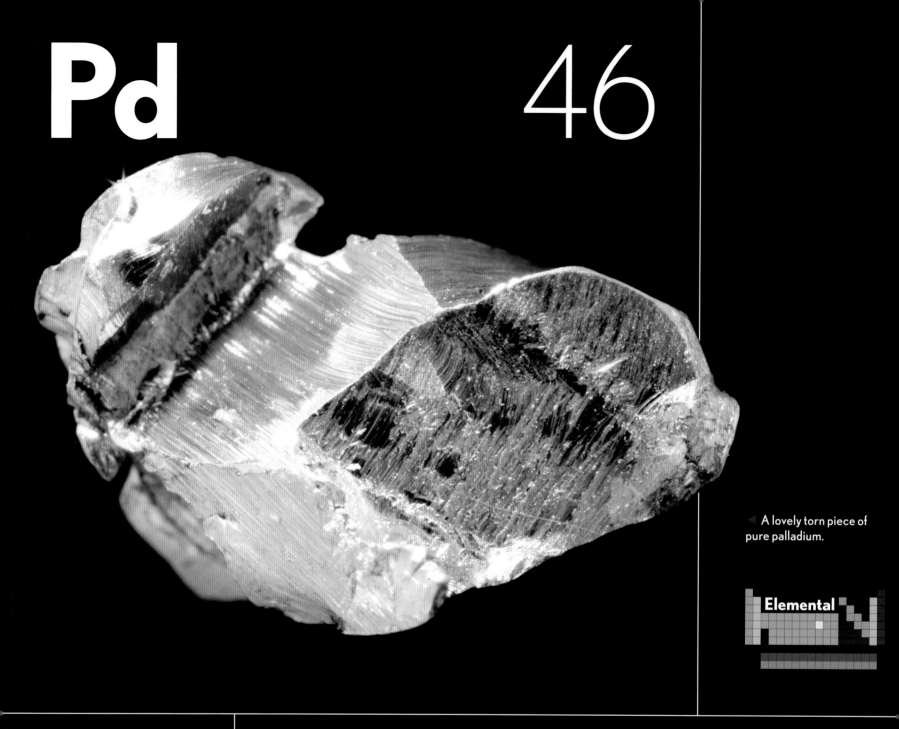

◀ A lovely torn piece of pure palladium.

Elemental

Palladium

PALLADIUM HAS AN astonishing ability to absorb hydrogen (1) gas. A solid chunk of palladium will absorb 900 times its volume in hydrogen! The gas virtually disappears in the solid metal, sneaking between palladium's crystal lattice of atoms. This would be an excellent way to store hydrogen in tanks for hydrogen-powered cars, if not for the fact that palladium is a precious metal nearly as expensive as gold (79).

Ag

▷ A tetradrachm (an ancient Greek silver coin) showing the name Alexander (as in the Great). Minted in 261 BCE, these coins are really, super-crazy old, yet there are a lot of them still around—no one ever throws away a coin.

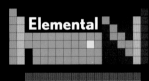

Elemental

SINCE ANCIENT TIMES, silver and gold (79) have been associated with glory and riches, but silver is definitely the junior partner. Historically, it has sold for about a twentieth of the price of gold, though in the last century the ratio has reached as much as a hundred to one. Gold is too expensive to use for making common coins, but for a long time, silver was just about right; it was used in common coins for almost 3,000 years.

Silver

Cd 48

A fish I cast out of solid cadmium, for no reason whatsoever.

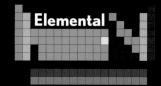

Cadmium

CADMIUM WAS PERHAPS best known for nickel-cadmium (ni-cad) batteries, but they have been replaced with lithium-ion batteries that are lighter, more powerful, and less toxic. Like lead (82) and mercury (80), cadmium can build up in the environment and in the human body, causing long-term damage. In brighter news, cadmium sulfide makes the classic pigment known as *cadmium yellow,* an intense color used by impressionist painters such as Claude Monet.

Pure indium is almost always sold in 1-kilogram bars (about 2.2 pounds). This is about half a bar. It is so soft that a bar like this can be cut in half with a knife (though it takes some effort).

Elemental

INDIUM IS NOT named for India or Indiana or any other place. It's named for the strong indigo-blue spectral line that was the first evidence of its existence. It's said that until 1924, only a gram of it had been isolated in the whole world, but these days hundreds of tons go into the production of LCD televisions and computer monitors every year.

Indium

Sn

50

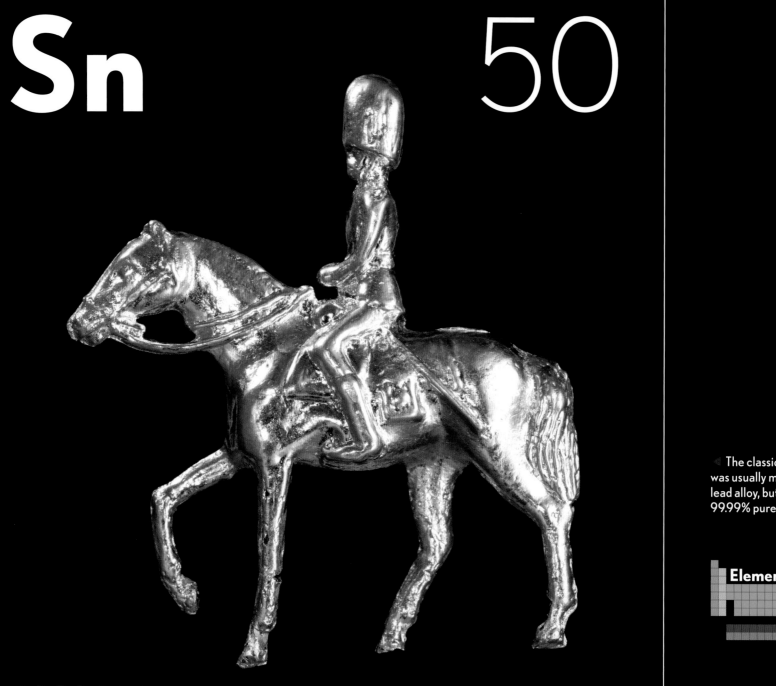

Elemental

Tin

TIN IS A lovely element. It's pretty much completely nontoxic, it stays shiny forever, it's easy to melt and cast into minutely detailed shapes, and it's not horribly expensive. Oddly, a lot of things called tin—including tin cans, tinfoil, and tin roofs—are not, and never wer▮ made of actual tin. The word has come to mean any sort of thin sheet metal.

Sb

Bulk antimony is sold in pretty lumps of crystal like this one.

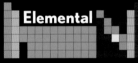

Elemental

Antimony

ANTIMONY IS A typical metalloid: clearly metallic in appearance, yet more brittle than ordinary metals. Adding antimony to lead (82) makes the lead harder. The right mixture of lead, tin (50), and antimony will expand just a tiny bit when it solidifies, which makes it wonderful for casting in molds. (Most metals shrink significantly when they go from a liquid to a solid, forcing people to take extra steps to get accurate, detailed castings without holes in them.)

Te 52

Beautiful crystals are forming on the surface of this disk of molten tellurium as it hardens.

Elemental

Tellurium

AN ODD PROPERTY of tellurium is that exposure to very small concentrations of it can cause you to smell like garlic for weeks! Despite this problem—and the fact that it is one of the rarest of all the elements—tellurium suboxide is used to coat DVD-RW and Blu-ray disks, and it's used in solar-panel cells and memory chips.

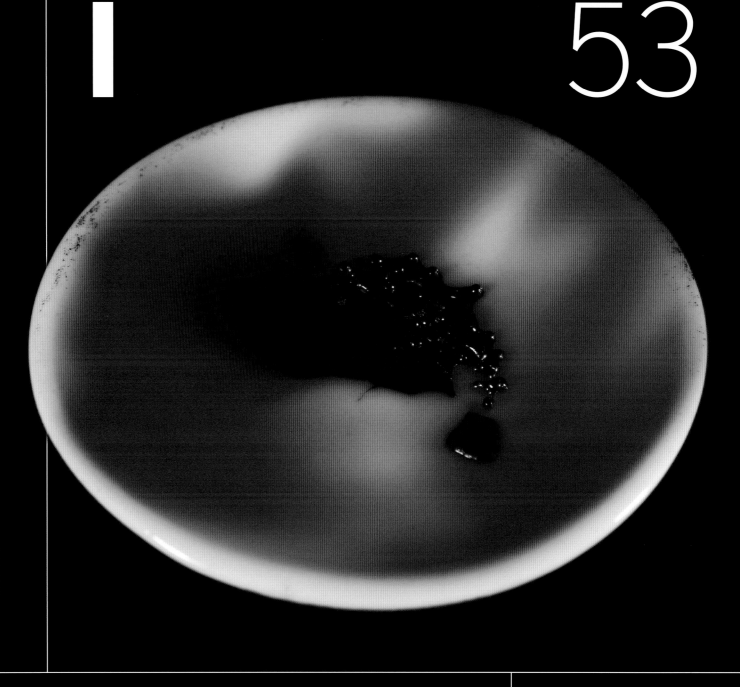

▷ Iodine evaporates into a wonderful violet vapor when heated. (There's a torch under the plate in this photograph.)

Elemental

Iodine

IODINE IS ONE of the most mellow elements in the halogen group. It was once widely used as a disinfectant. It's a solid at room temperature, but just barely. Gentle heating will melt it, at which point iodine immediately evaporates into a thick, dense, beautiful, dangerous purple vapor.

Xe

54

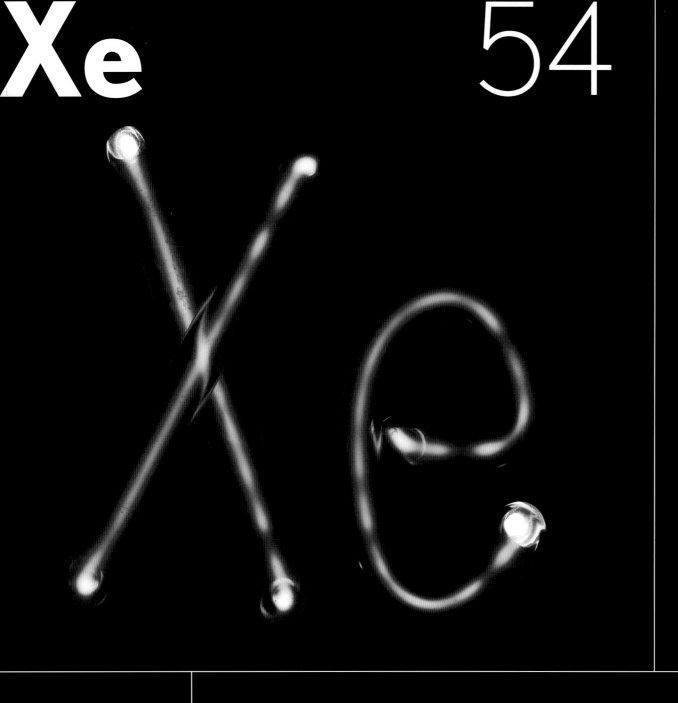

The xenon gas in this tube is being excited by electricity, which creates the lovely pale-violet glow.

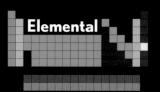
Elemental

Xenon

XENON GAS HAS low thermal conductivity, meaning it conducts heat slowly. This makes it useful for IMAX movie projectors, which use fantastically bright 15-kilowatt xenon short-arc lamps to create their huge projected images.

The bulbs are filled with xenon at such a high pressure that they must be stored in special protective enclosures and handled by people dressed in protective clothing, because of the risk of explosion.

Cs

55

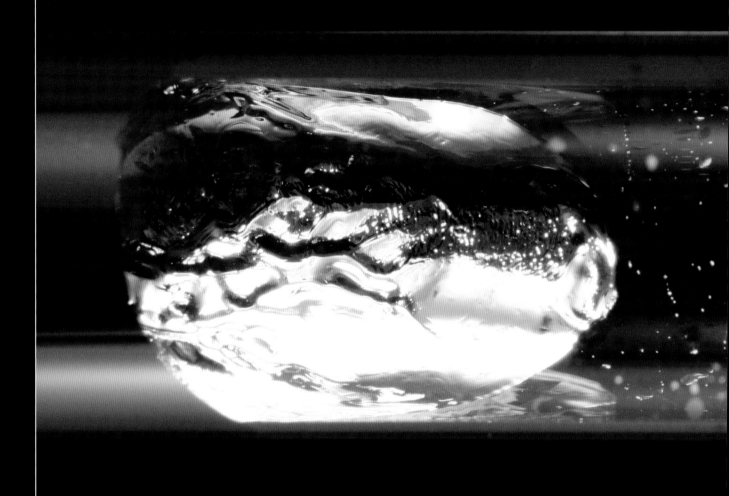

▷ The cesium in this ampoule melts if you hold it in your hand for even a minute, turning into the prettiest gold liquid. If the ampoule were to break in your hand, the resulting fire would not be nearly as nice.

CESIUM IS THE most reactive of all the alkali metals. When you drop a piece into a bowl of water, it explodes instantly, sending water flying in all directions. But despite being technically the most reactive, another alkali metal, sodium (11), gives more reliable, bigger explosions. This makes sodium the preferred metal to throw into lakes. Where cesium shines is in the construction of the most accurate clocks in the world, known as cesium fountain clocks.

Cesium

Ba 56

This mineral barite (barium sulfate) came from a mine in Huancavelica, Peru.

Elemental

Barium

THE NAME *BARIUM* comes from the Greek for "heavy." While barium in its pure form is not particularly heavy—it's actually less dense than titanium (22), a metal known for being light—many of its compounds are usefully dense. Barium sulfate is opaque to x-rays, so it's given to patients as a dye they can drink to help outline various parts of their digestive system in x-ray images.

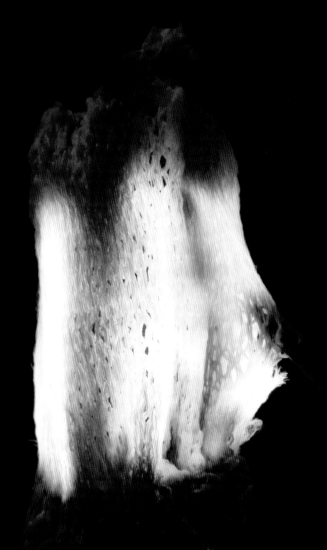

A large torn ingot (metal cast into a shape that's easy to work with) of pure lanthanum metal.

Elemental

LANTHANUM IS THE first of the rare earth elements known as the *lanthanides*, the uppermost of the two rows of elements commonly shown outside the main periodic table. Lanthanum is one of the most abundant of the rare earths (which really aren't that rare). Because it likes to catch fire at fairly low temperatures, it's often mixed with iron (26) and used in lighter "flints" that create sparks when vigorously scratched by a thumb wheel.

Lanthanum

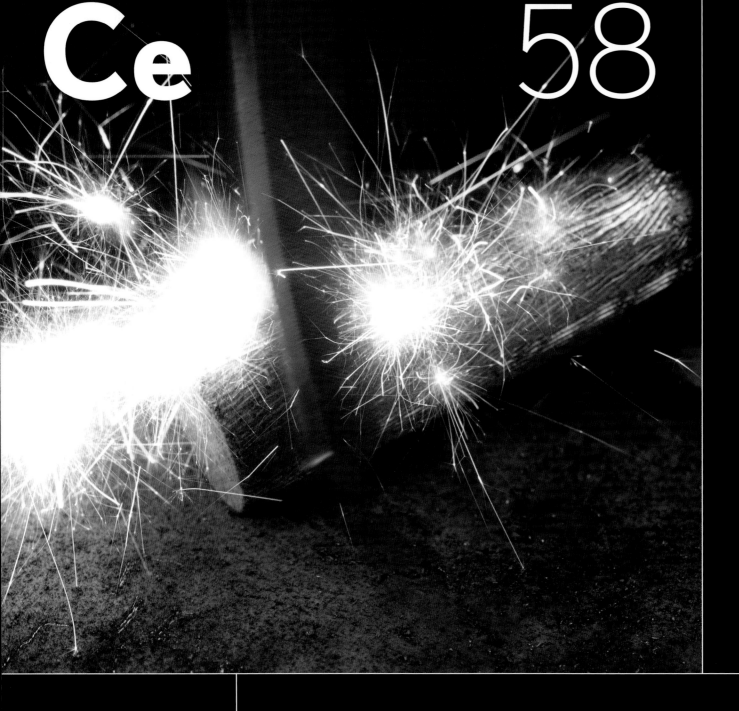

Ce 58

This huge rod of cerium-lanthanum-iron alloy is half an inch in diameter. It's basically a giant lighter flint that creates a shower of sparks when you scrape it with a steel blade.

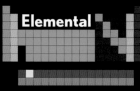

Cerium

CERIUM METAL IS pyrophoric, meaning that it can catch fire when you scratch, file, or grind it. This doesn't mean the whole lump catches fire. Instead, the shavings burn as they are formed, making the material very sparkly. Not surprisingly, this makes cerium useful in lighter flints.

Pr

59

▷ A block of pure praseodymium that has oxidized a little bit.

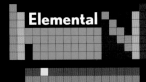

A UNIQUE USE of praseodymium is in the production of the "didymium" eyeglasses worn by glassblowers to protect their eyes from the very bright sodium (11) emission lines given off by hot soda-lime glass. Praseodymium blocks the yellow light, so glassblowers can stare directly at the torch that is heating the glass and see nothing but the dull blue glow of the torch flame. Remove the glasses, and the blinding yellow light forces them to look away.

Praseodymium

Nd

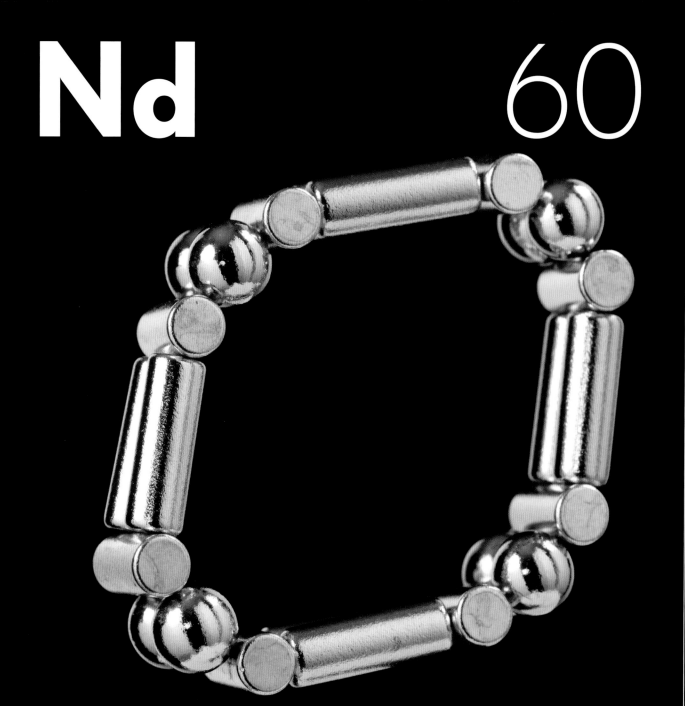

◁ A chain of neodymium magnets is strong enough to wear as a bracelet without a string connecting any of the beads.

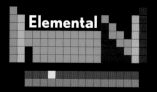

Neodymium

NEODYMIUM MAGNETS, WHICH are actually made of a neodymium-iron-boron alloy, are by far the strongest readily available permanent magnets. They are so strong that they are dangerous to be around, especially if you have more than one. They can jump toward each other from a foot or more away. Heaven help you if your fingers get between them!

▷ This glowing promethium button was created using leftover materials from making diving watches.

Elemental

Promethium

PROMETHIUM IS ONE of two elements below bismuth (83) in the periodic table that are unstable. The other is technetium (43). Promethium was, at one time, mixed with zinc sulfide to create a phosphorescent material that was used to create glowing dials and markings for things like compasses. Few examples of these devices survive, and none of them work anymore. Since then, few new applications have been found for this element.

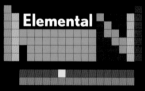

A coin made of pure samarium is one of a series of coins created from nearly every practical element.

Elemental

Samarium

SAMARIUM IS NAMED for the mineral samarskite, which in turn is named for its discoverer, Vasili Samarsky-Bykhovets of Russia. A case could be made that, along with seaborgium (106) and oganesson (118), samarium is the only other element named after a living person. Unlike seaborgium, however, samarium does not directly honor the person. Indirect naming through a previously named mineral doesn't count in my book (and this, you will note, is my book).

Eu

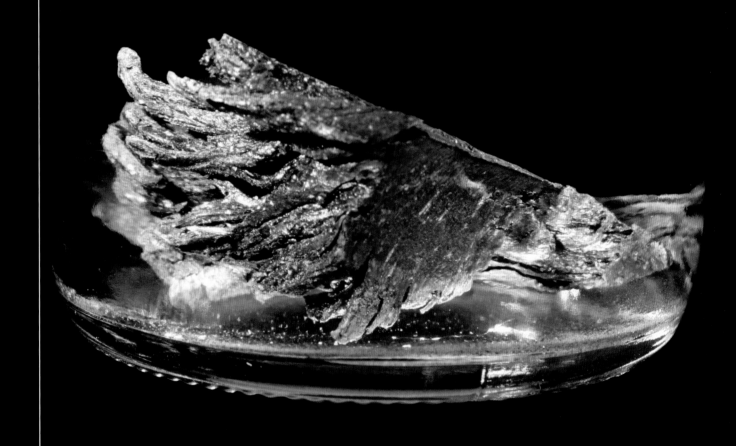

Pure europium oxidizes over time, even when it's stored under oil.

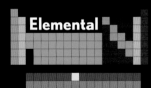

Elemental

EUROPIUM IS NAMED for the continent of Europe. Its applications are somewhat unusual for a rare earth element as they center not on magnetism, but rather on luminosity (how brightly it glows). For example, europium is used to make phosphorescent paint, including some amazing modern varieties that can glow brightly for many minutes, or dimly for many hours, after being exposed just briefly to a strong light source.

Europium

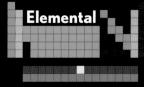

Elemental

Gadolinium

ONE OF THE main uses of gadolinium is as a contrast agent for MRI (magnetic resonance imaging) scans. An MRI uses a strong magnetic field and radio frequency probe to create an image of where particular atoms are located in the body.

If you inject a person with gadolinium, you can easily track where it goes because of gadolinium's unusual magnetic properties. This lets you find, for example, clusters of blood vessels where they don't belong, which can indicate a tumor.

Tb

65

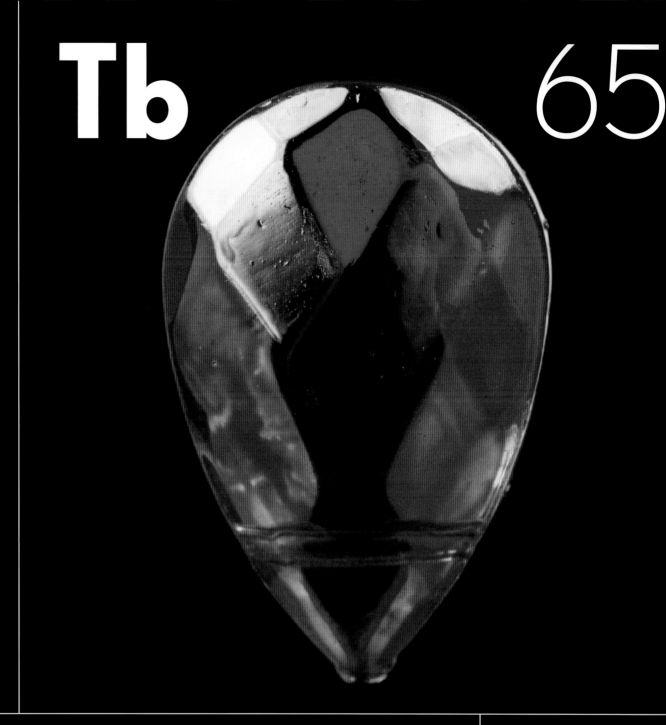

A terbium-enriched, red-glass teardrop made because it looks cool.

Terbium

TERBIUM CAN CHANGE shape when placed in a magnetic field. A rod of terbium alloy called *terfenol* will instantly grow longer or shorter depending upon the strength of the field. If something gets in the way of it while it's trying to change in length, it will push with great force. This lets a terfenol rod with an audio amplifier turn almost any solid object, such as a window or tabletop, into a speaker—the rod is strong enough to shake the whole object.

Dy

66

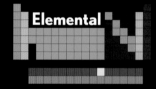

◁ Dendritic (a shape that looks like the branches of a tree) crystals of pure dysprosium.

Elemental

Dysprosium

IT'S NOT THAT you can't find anything useful about dysprosium, but it tends to live up to its name, which is from the Greek *dysprositos*, meaning "hard to get at." For example, dysprosium iodide and bromide are used to add spectral lines to the red color range in high-intensity discharge lamps. That took me a while to find out. In a few years, when these lights are obsolete, I'll have to find a whole new fact.

Ho

The polycrystalline surface of pure holmium metal.

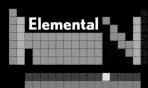

Elemental

HOLMIUM WINS THE prize among all the rare earth elements for having the highest value of a particular magnetic property known as the *magnetic moment*. This means that when holmium is placed in a magnetic field, the holmium atoms line up with the field and concentrate it, bringing the magnetic lines of force closer together. This makes the field more intense. In other words, a block of holmium can turn a very strong magnetic field into a crazy-strong one.

Holmium

Er

68

◁ Erbium impurities create the pink color in these pretty glass rods.

Elemental

Erbium

ERBIUM HELPS TO amplify pulses of light without having to convert them into electrical signals like you have to do for lightbulbs and LEDs. This property is great for optical amplifiers in high-speed, long-distance fiber-optic data networks. A pulse of laser light coming down an optical fiber is "pumped" into a section of the fiber that has been enriched (or "doped") with erbium. When this data pulse passes through, it comes out the other side much stronger than when it entered.

Tm

69

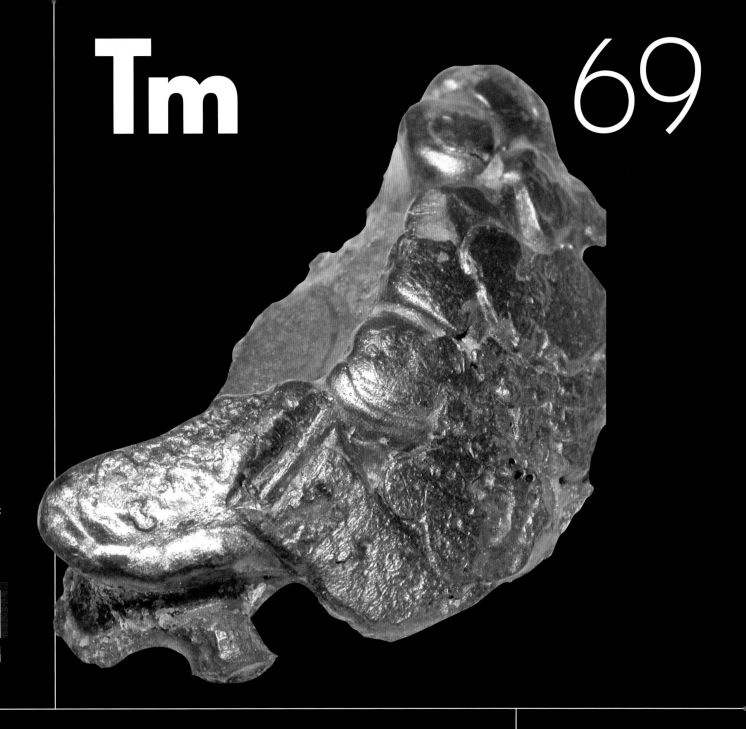

A large, melted lump of thulium metal.

Elemental

THOUGH MOST PEOPLE have never heard of thulium, lighting designers love it. High-intensity arc lights are made with a mixture of elements to build up the spectrum—the color—of the light they make. Thulium's main purpose in life is to provide a broad range of green emission lines in an area of the spectrum not readily covered by any other element. There are people in the lighting industry who will fight you if you dis thulium, as I once made the mistake of doing.

Thulium

Yb

The mineral xenotime is mostly made up of ytterbium.

Elemental

Ytterbium

YTTERBIUM IS ONE of four elements—along with yttrium (39), terbium (65), and erbium (68)—that are named after the town of Ytterby in Sweden. It's mainly used in lasers.

Lu

71

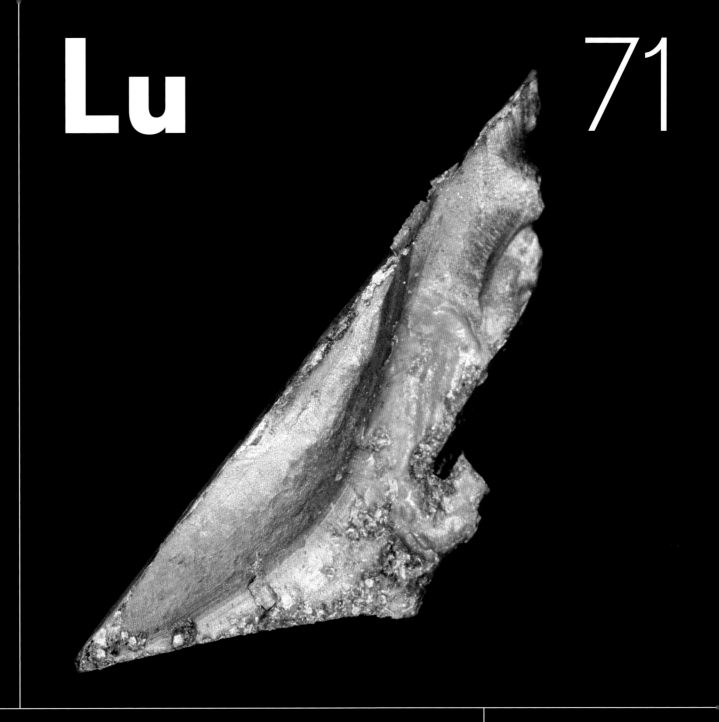

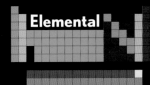

▷ A cut shape of pure lutetium.

Elemental

THERE AREN'T VERY many interesting things to say about lutetium that we haven't already covered with all the other rare earth elements, except that it was the last natural rare earth element to be discovered. It's also among the hardest of all the lanthanides. Because there's so little commercial application for lutetium, I wouldn't be surprised if the largest market for it is element collectors like me!

Lutetium

Hf

A high-purity crystal of hafnium.

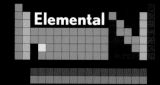

Hafnium

HAFNIUM HAS A very high melting point and is extremely resistant to corrosion, even at very high temperatures. This makes it an excellent element to use to create, say, a hafnium carbide cutting bit, which can be used to cut through steel.

Ta

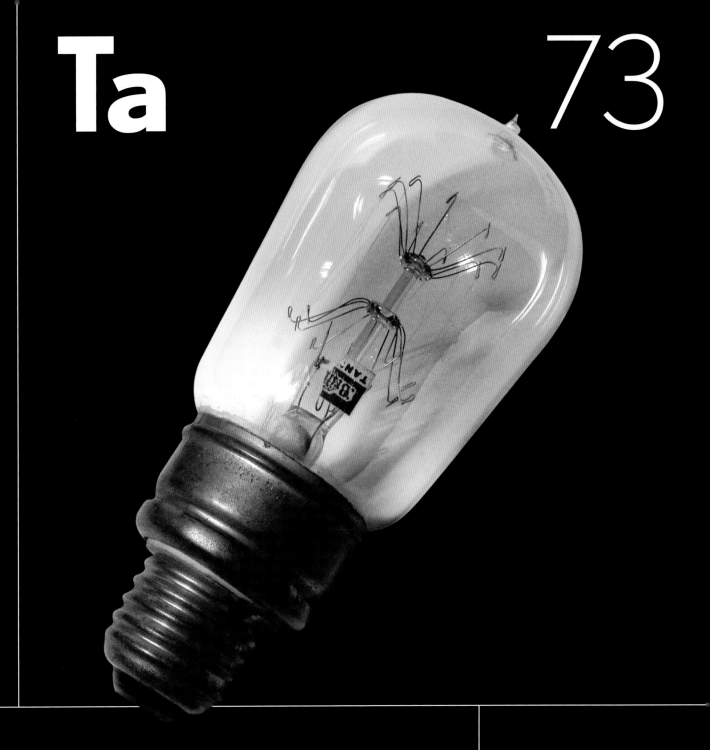

▶ An antique tantalum-filament lightbulb.

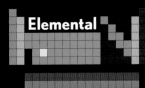

TANTALUM IS NOT limited in its applications, just in name recognition. In fact, it plays a critical role in dampening high-frequency electrical noise in all kinds of electronic devices, from cell phones and computers to medical equipment and talking dolls. Historically, tantalum-filament lightbulbs were used on the *Titanic* because they were so much more reliable than older carbon-filament bulbs and could be left on at night.

Tantalum

A filament from a tungsten incandescent lamp, which is now almost a relic of the past.

Elemental

Tungsten

THE METALS FROM tungsten through gold are all very dense, but tungsten is by far the cheapest. It's good when you want to put a lot of weight in a small space. For example, you can get tungsten throwing darts and tungsten blocks for knocking dents out of your car. Before the 1990s, nearly all lightbulbs had a tungsten filament inside. But these bulbs waste about 90% of the electricity they use, and thankfully nearly all of them have now been replaced with fluorescent or LED bulbs.

Re

75

Elemental

Rhenium

RHENIUM WAS THE last stable element to be discovered. Most rhenium is used in nickel-iron superalloys for turbine blades in fighter-jet engines.

Even though there are not that many fighter jets being made, they consume a whopping three-quarters of the world's annual rhenium production.

Os 76

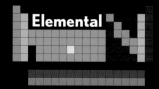

Osmium tetroxide crystals are toxic and must be kept in a sealed glass ampoule.

Osmium

(Elemental logo)

OSMIUM, LIKE COPPER (29) and gold (79), is a metal that is not gray or silver in color. Even though it *looks* like just another silvery metal, it's not. Its slight bluish tint is so faint you have to work to convince yourself you're really seeing it, but it's there. Osmium is the hardest metal element! That, by the way, does not mean it's the hardest material, or even the hardest element, just the hardest pure metal. It's also the densest of all the elements.

Ir

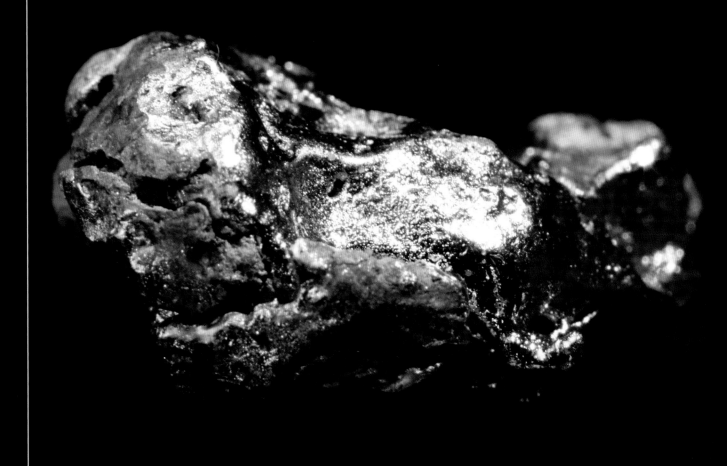

▷ Iridium is extremely hard to melt. I could only get this lump to melt about halfway, hence its odd shape.

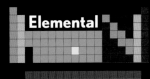
Elemental

ALTHOUGH OSMIUM (76) is the densest element, it only wins that title by less than one-tenth of 1%; in *very* close second place is iridium. Iridium is really expensive, so it's mostly used in places where you need only a very small amount. For example, some high-grade automobile spark plugs are equipped with tiny iridium tips that last up to 100,000 miles, far more than the tips of conventional plugs.

Iridium

Pt

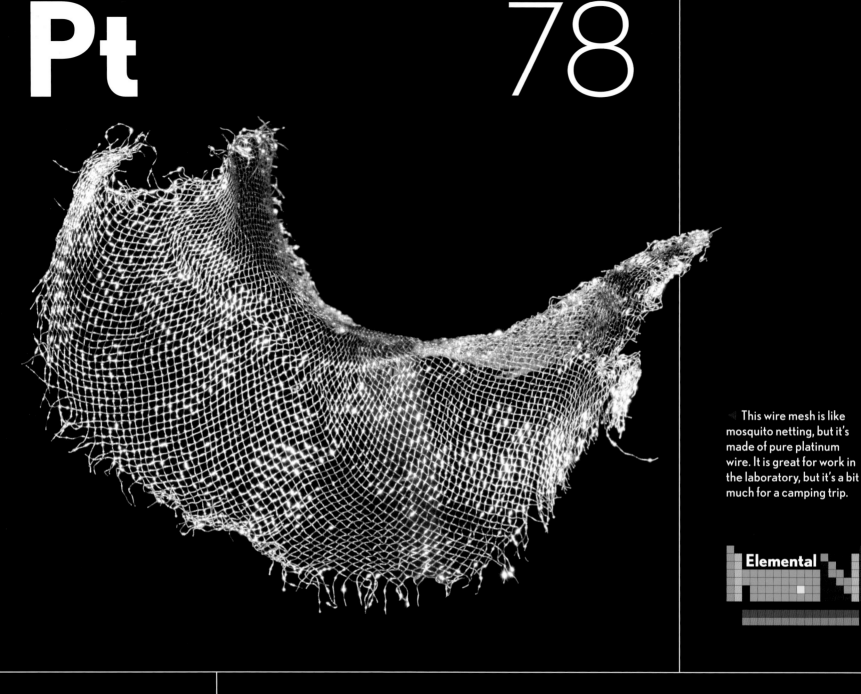

This wire mesh is like mosquito netting, but it's made of pure platinum wire. It is great for work in the laboratory, but it's a bit much for a camping trip.

Elemental

Platinum

PLATINUM IS *THE most* prestigious element. Sure, gold (79) is great, but platinum is always better. Platinum is more abundant in Earth's crust than other similar metals, but it is significantly more expensive because the demand

is so high. More than any other metal, platinum is able to withstand powerful acids and high temperatures—almost anything you can throw at it—without so much as staining.

Au

▷ This gold is only about 500 atoms thick, and it's so fragile that it can only be picked up with static electricity using something like the bristles of a hairbrush.

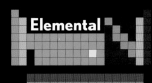
Elemental

Gold

GOLD IS THE gold standard of metals. It is inherently valuable, and there is very little of it around. All the gold ever mined in the history of the human race would fit into a cube about 60 feet on edge. Gold is undeniably beautiful, and it is the only metal that both is colored and keeps its shine and beauty forever. The saying "diamonds are forever" should actually be "gold is forever," because, unlike gold, diamonds are easily destroyed by heat.

Hg

A pool of mercury that I carefully lit and lovingly photographed.

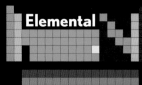

Mercury

MERCURY IS MAGICAL. It's so dense that if you try to take a bath in it, you sink in just a few inches. People do this, but it's not a good idea. For thousands of years, mercury was treated as a wonderful thing to play with, but all the while it was damaging people's central nervous systems, making them go crazy. Mercury is the worst kind of poison—the kind you don't notice until years after the damage has been done.

▷ This large piece of thallium metal is kept in a safe because it could potentially poison hundreds of people.

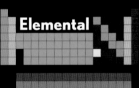

Elemental

Thallium

THE TRICK TO poisoning someone is finding a new kind of poison with symptoms no one recognizes. A hundred years ago, that was thallium. Like arsenic (33), thallium is a seriously toxic element, but it's much less popular in murder mystery novels. If you'd like to check whether you are the victim of a thallium poisoning, the symptoms include vomiting, hair loss, confusion, blindness, and stomach pain—each of which, you'll notice, is also a symptom of a hundred other things.

Pb 82

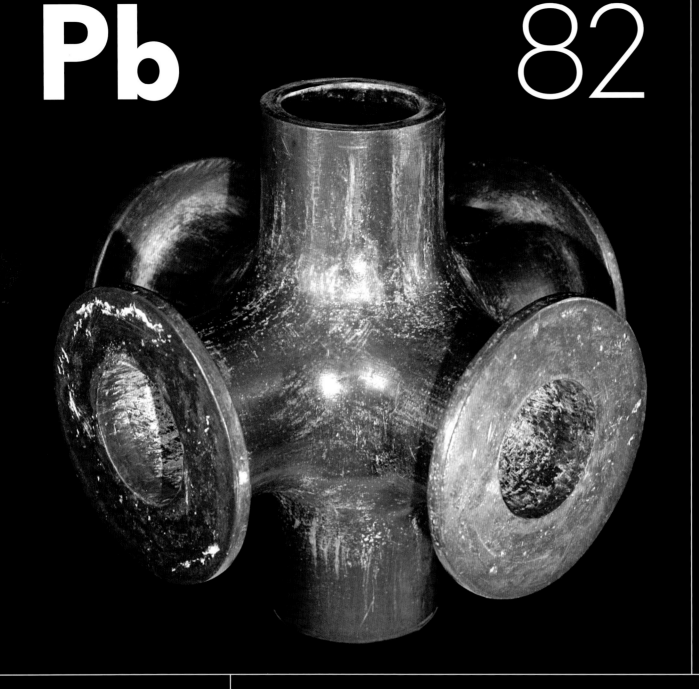

This six-way union was hammered out of a lead sheet by an apprentice pipefitter, which really impressed the boss.

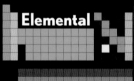

Elemental

Lead

AS LITTLE AS 2 grams of lead is a lethal dose—when it's delivered from the barrel of a gun. Lead is the preferred metal for bullets, because it's quite dense, allowing a lot of mass to fit into a small space.

Just like the last few elements we've discussed, lead is toxic, and along with mercury (80), it has been responsible for terrible environmental contamination in the modern age.

Bi

83

Bismuth spontaneously forms large "hopper" crystals when it cools. When very pure bismuth is cooled very slowly, these can grow to huge sizes. This one is more than 4 inches tall.

THE ACTIVE INGREDIENT in Pepto-Bismol brand upset-stomach medicine is 57% bismuth by weight. This is really quite odd when you consider the fact that bismuth sits smack in the middle of the toxic heavy metals lead (82) and polonium (84). But, so far as we know, the metal form of bismuth is completely nontoxic.

Bismuth

Po 84

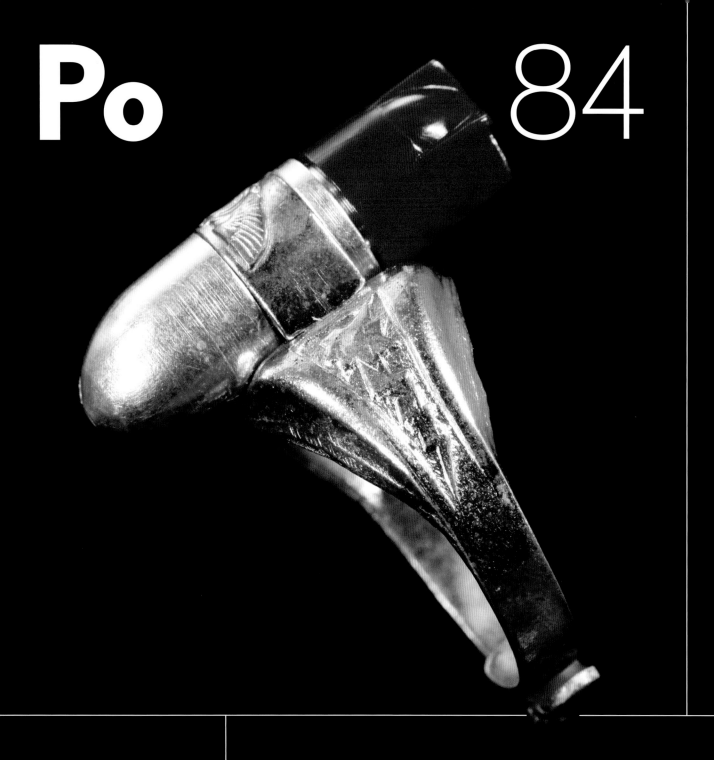

This Lone Ranger Atomic Bomb Spinthariscope Ring, with a trace of radioactive polonium inside to make it sparkle in the dark, was a toy offered by Kix cereal in 1947—proof of how differently people used to think about atomic bombs.

Polonium

POLONIUM WAS DISCOVERED by Marie and Pierre Curie and named for Marie's native home of Poland. The most common modern application of polonium is in antistatic brushes. These brushes are used on phonograph records and film negatives to get rid of static charges that attract dust. Oh wait, no one uses phonograph records or film negatives anymore, so I guess we have to count those as historical applications as well!

At

85

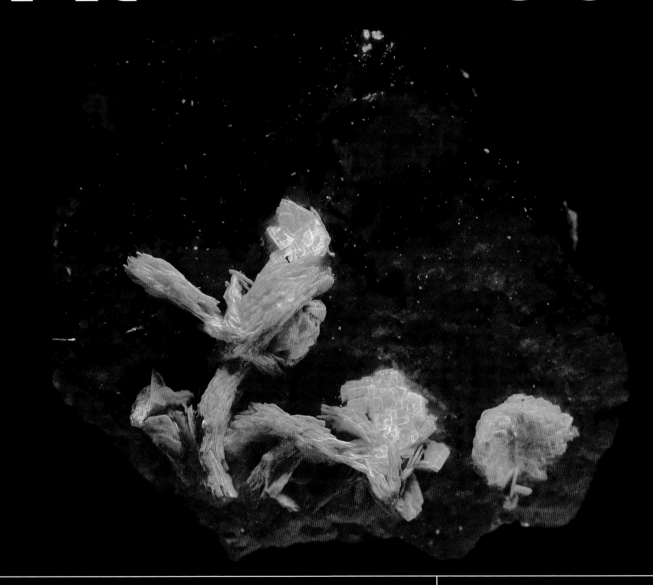

▷ This beautiful fluorescent uranium mineral called *autunite* may or may not contain an atom of astatine at any given moment in time.

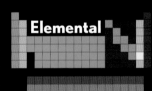

Elemental

Astatine

ASTATINE IS REALLY frustrating for element collectors. Although it is considered naturally occurring, its half-life (the time it takes for half of it to vanish through radioactive decay) is only 8.3 hours, which means that whenever astatine occurs naturally, it doesn't stick around for very long. Despite its short half-life, astatine is being studied for use in radiation therapy for cancer.

Rn

86

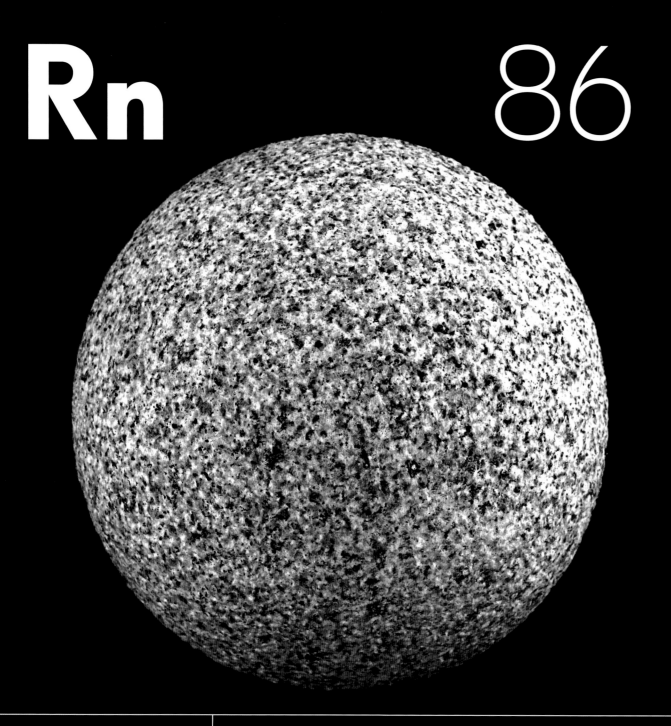

This granite ball represents the major source of radon: traces of uranium and thorium decaying in bedrock. Granite buildings, famously including Grand Central Station in New York City, are significantly radioactive because of these traces.

Elemental

Radon

RADON IS A heavy radioactive gas with a half-life of only 3.2 days. Despite its short half-life, there is quite a lot of it around because it is generated by the decay of uranium (92) and thorium (90), which exist in large quantities in granite bedrock. Granite buildings emit significant amounts of radiation—Grand Central Station in New York City is famously radioactive for this reason.

Fr

87

This piece of the mineral thorite might contain an atom of francium if you look closely.

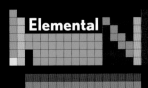

Elemental

FRANCIUM IS THE least-stable naturally occurring element and the last element discovered in nature. It was first found in, you guessed it, France. With a half-life of only 22 minutes, there are no commercial applications for it.

Francium

Ra

88

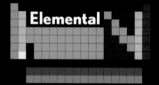

Radium

RADIUM WAS THE titanium (22) of the early 1900s. It was the fabulous element everyone wanted to associate their products with. The best-known use of radium was to create glowing watch hands. Women painted these by hand with tiny brushes they sharpened by licking them. This was a bad thing considering the paint was radioactive! The women became sick and many died, which finally convinced people something had to be done about worker safety in factories.

Ac

This sample of vicanite—from the Vico volcanic complex in Tre Croci, Italy—probably doesn't have any actinium in it right now, but once in a while it might have an atom or two.

Elemental

ACTINIUM IS THE first of the actinide series of the rare earths, the ones placed in the very bottom row of a standard periodic table. Actinium is so radioactive that it glows on its own without a phosphorescent screen (which you need to see the glow from less radioactive elements). While there are some experimental applications of actinium, very little of it is made or used.

Actinium

Th
90

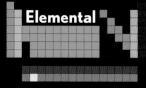
Chips of pure thorium metal.

Elemental

Thorium

THERE IS MORE thorium in Earth's crust than there is tin (50)—almost three times as much. Because there is so much thorium, it was used for many years purely for its chemical properties even though it is radioactive. Thorium oxide, for example, was used until recently in camping lanterns, where it glowed brightly when heated by a gas flame.

Torbernite is a lovely green uranium mineral I chose to represent protactinium out of sheer desperation. There is no practical way to get or photograph actual protactinium, but some atoms of it might be in this rock from time to time.

Elemental

Protactinium

PROTACTINIUM IS ANOTHER element that seriously annoys element collectors. Its half-life is quite long— 32,788 years. A large lump, while dangerously radioactive, would be entirely fine to show off in a nice, lead-lined display case. But the fact that it's entirely unavailable except in tiny amounts for research purposes means that you can never find it on eBay.

U 92

Pure uranium metal is perfectly legal to own (up to 15 pounds at any one time), and there are actually a few companies that sell it to element collectors. This 30-gram piece came from one such company.

Uranium

IT IS IMPOSSIBLE to talk about uranium without acknowledging that the first atomic weapon used in anger was a uranium fission bomb. It was built in secret, deep in the desert of New Mexico, and detonated over the unsuspecting city of Hiroshima, Japan, in 1945. Uranium was also used for many years to make an orange glaze for Fiestaware dishes, where it harms people by radiation and by heavy metal poisoning—which doesn't stop some people from using these dishes.

Np

Some aeschynite from Molland in Iveland, Norway. It doesn't really have any neptunium in it, but it's radioactive and you *could* get actual neptunium from it.

Elemental

NEPTUNIUM WAS THE first transuranic element (an element after uranium on the periodic table) to be discovered. It was found in 1940 by scientists at the University of California, Berkeley. Uranium is usually considered the last naturally occurring element, but in fact scientists believe that very tiny amounts of neptunium should exist in uranium-bearing minerals because of the nuclear side reactions that happen when uranium decays.

Neptunium

Pu

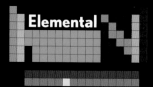

This plutonium heart pacemaker battery case is empty—fortunately. If it were full, possession of it anywhere outside a body would be a crime.

Plutonium

PLUTONIUM IS OFTEN called the most poisonous element. Private ownership of plutonium is absolutely forbidden, with one tiny exception. Pacemakers today use lithium batteries, but a few people— no one knows exactly how many—still have old models powered by plutonium thermoelectric batteries. When one of these people dies, their pacemaker is returned to Los Alamos National Laboratory, where the plutonium in it can be loved and cared for.

Am

▷ The radioactive americium button from inside a common ionization-type smoke detector. Underneath the gold foil is 0.9 microcurie of the element (a super small amount).

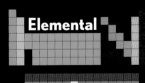

Elemental

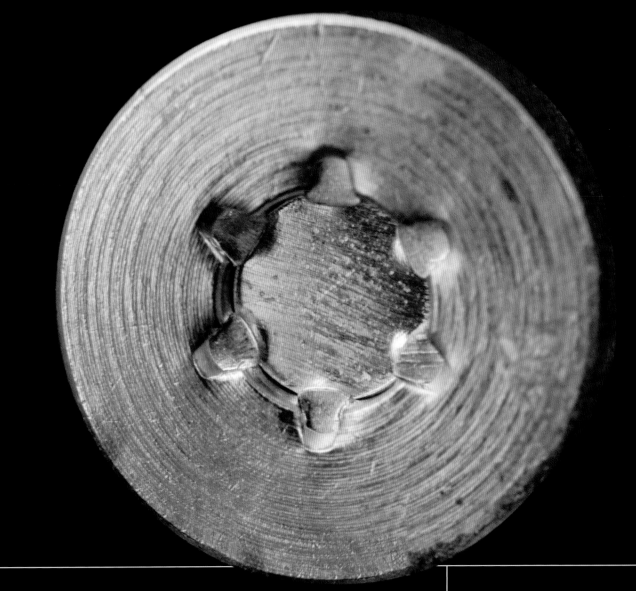

AMERICIUM IS SIGNIFICANTLY more radioactive than weapons-grade plutonium (94) and at least as toxic. Yet you can buy some in any hardware store or supermarket. Nearly all smoke detectors contain a tiny trace of americium (don't worry, the radiation can't get out of the detector). With americium, we reach the end of the line for element collectors. It is the very last element that is legal to own without expensive special licenses.

Americium

Cm 96

Marie Curie, for whom curium is named.

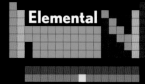
Elemental

Curium

CURIUM, CURIOUSLY, WAS not discovered by the dynamic duo of Marie and Pierre Curie. They just discovered polonium (84) and radium (88). In fact, none of the elements named for people were discovered by those people. Curium, for example, was discovered by a large team led by Glenn T. Seaborg, Ralph A. James, and Albert Ghiorso at the University of California, Berkeley.

Bk

The great seal of the University of California, Berkeley, where Glenn T. Seaborg discovered berkelium and many other elements.

Elemental

THE LONGEST HALF-LIFE of any berkelium isotope is 1,379 years. This means if you had a 1-pound block of berkelium and let it sit for 1,379 years, you would have only half a pound of berkelium left. If you let it sit for another 1,379 years, you would have a quarter of a pound of berkelium, and so on. Virtually no practical applications have been found for berkelium.

Berkelium

Cf 98

The great seal of the state of California, for which californium is named

Elemental

Californium

CALIFORNIUM IS THE last element that has any application whatsoever. It can be used as an extremely powerful neutron emitter. Because neutrons carry no charge, they are not repelled by negatively charged electrons or by positively charged protons. This means californium's neutrons can pass through perfectly ordinary solid matter with ease and turn it radioactive.

Es

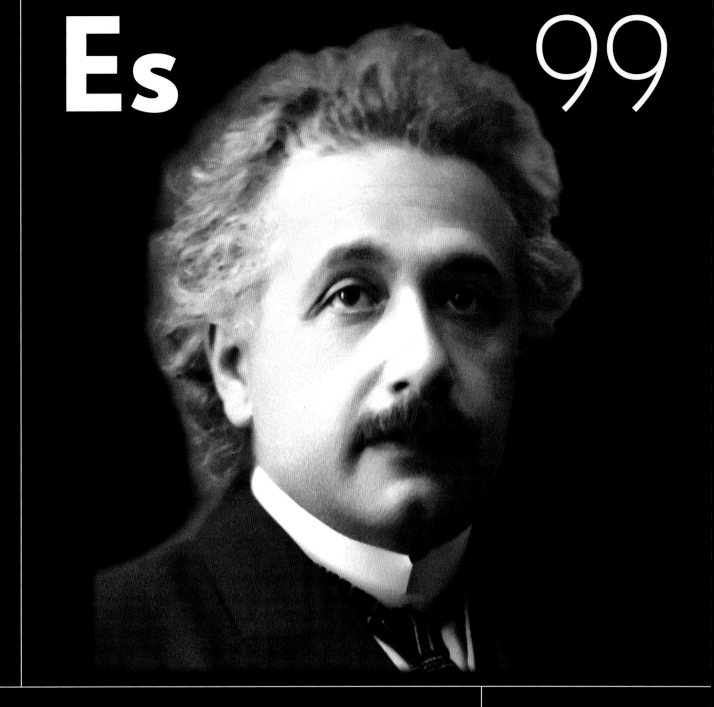

▷ Albert Einstein is the most famous scientist of all time, so it makes sense that an element is named after him.

Elemental

GETTING AN ELEMENT named after yourself is not easy. Winning a Nobel Prize is no big deal by comparison. There are more than 800 Nobel Prize winners, with more added every year, but just a handful of people are, or can ever be, honored with an element. Albert Einstein, however, was a shoo-in. He was the most famous scientist in history while he was alive, and he may just be the poster boy for super-smartness forever.

Einsteinium

Fm

100

Fermium

FERMIUM IS NAMED after Enrico Fermi, creator of the world's first nuclear reactor, known as the *Chicago Pile-1*. Of course, none of this has anything to do with the element fermium, which (just like the remaining 18 elements) has no applications.

ELEMENTS 101 THROUGH 109 are the range of elements where things go from "it has no applications, but at least a visible amount of it has been created" to "you can list exactly how many atoms have been created and when." By the time we reach the last one, meitnerium, we're talking about less than 24 atoms total. Nuclei in this part of the table are too big and too unwieldy to hang out together for more than a few hours. The longest lived is mendelevium (101), with a half-life of 74 days, followed by rutherfordium (104), whose half-life clocks in at only 19 hours. The award for shortest half-life span goes to meitnerium (109) at just 43 minutes.

NEARLY ALL THE transuranic elements were named in honor of someone. Most of these people won Nobel Prizes, but not all of them. Dmitri Mendeleev did not win a Nobel Prize because it didn't yet exist when he invented the periodic table.

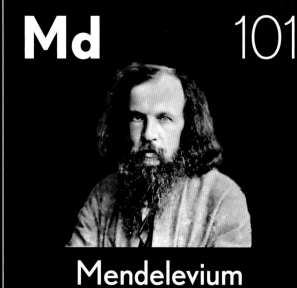

Md 101

Mendelevium

ALFRED NOBEL DIDN'T win a Nobel Prize because he invented Nobel Prizes.

No 102

Nobelium

LAWRENCIUM WAS A natural choice for the name of this element since Ernest Lawrence built the first working cyclotrons, the machines used to discover many of the elements in this range.

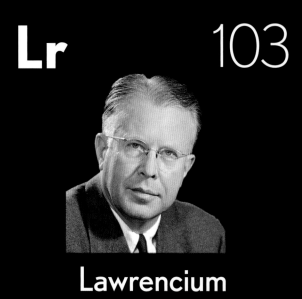

Lr 103

Lawrencium

Rf 104

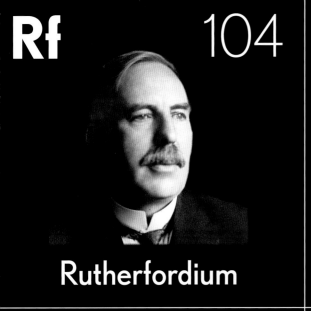

Rutherfordium

ERNEST RUTHERFORD DATES back even further than Lawrence. He was the first person to discover that elements had nuclei.

Db 105

Dubnium

THERE WAS A lot of arguing about what to name this element, but people finally agreed on *dubnium* in 1997. The name comes from the town of Dubna, the site of the Soviet Joint Institute for Nuclear Research (JINR), where scientists discovered the element in 1968.

Sg 106

Seaborgium

SEABORGIUM IS NAMED after the American nuclear chemist Glenn T. Seaborg, who is also on the list of people credited with the discovery of californium (98), plutonium (94), americium (95), curium (96), berkelium (97), einsteinium (99), fermium (100), mendelevium (101), nobelium (102), and, of course, seaborgium.

Bh 107

NIELS BOHR WON the Nobel Prize for Physics in 1922 for his important work in describing what's inside an atom, known today as the Bohr model.

Bohrium

Hs 108

THE NAME *HASSIUM* comes from the German state of Hesse, where this element was discovered. It's the German equivalent of californium (98).

Hassium

Mt 109

LISE MEITNER DIDN'T win a Nobel Prize mainly because she was a woman, but she did get the last laugh. Many people believe she deserved to share the 1944 Nobel Prize in Physics with Otto Hahn for the discovery of nuclear fission, but a Nobel Prize is a cheap trinket in comparison to having an element named in your honor.

Meitnerium

WE HAVE NOW reached a range of elements that, even though they have all been discovered, don't actually exist. What I mean is that no atoms of any of these elements are known to exist on Earth, unless, as you are reading this, someone happens to have their heavy-ion research accelerator switched on to make some.

Ds 110

Darmstadtium

DARMSTADTIUM IS NAMED for the German city of Darmstadt, home of the Gesellschaft für Schwerionenforschung (Institute for Heavy Ion Research).

Rg 111

Roentgenium

WILHELM CONRAD RÖNTGEN discovered x-rays, which makes it somewhat ironic that the element named after him does not emit x-rays when it decays moments after being created.

Cn 112

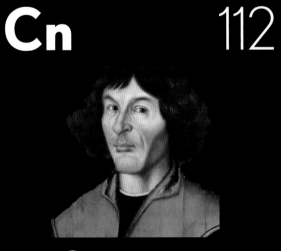

Copernicium

COPERNICIUM, DISCOVERED IN 1996 but not officially named until 2010, has the honor of being the only element (other than nobelium [102]) to be named for someone who didn't have much to do with chemistry or nuclear physics. Nicolaus Copernicus's main claim to fame is that he was a great astronomer.

FI 114

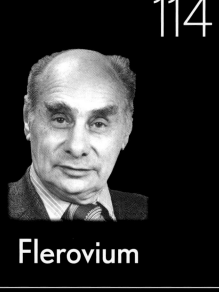

Flerovium

FLEROVIUM, KNOWN UNTIL May 2012 as ununquadium, was discovered in 1998 by a team at the Joint Institute for Nuclear Research in Dubna, Russia. It was—eventually—named for the founder of that institute, physicist Georgy Flyorov.

Lv 116

Livermorium

LIVERMORIUM IS NOT actually made of liver. It's named after the Lawrence Livermore National Laboratory, which is named after Robert Livermore. Mr. Livermore is unusual when it comes to people who have elements named after them (even indirectly named after them), because he wasn't a scientist. He was a cattle rancher, but the laboratory was built on his land and so named after him. Then the element was named after the lab. That's how you get this weird naming history.

IN DECEMBER 2016, the powers that be officially named the very last four elements: Nihonium (113) comes from *Nihon* (the Japanese name for "Japan"). Moscovium (115) was named for Moscow, and tennessine (117) was named after Tennessee. Oganesson (118) was named for scientist Yuri Oganessian, who is only the second person ever to have an element named after them while they were still alive.

In 2016, we reached a real milestone. Every single one of the 118 elements in a standard periodic table finally got a permanent name! We're done! No more need for new versions of the periodic table, ever. Maybe.

There is no real reason for the elements to stop at 118. That's just the last one that fits into the standard arrangement of the periodic table. Since none higher has been discovered, there's no reason to add a whole new row. Yet.

Nh 113

Nihonium

Mc 115

Moscovium

Ts 117

Tennessine

Og 118

Oganesson

The Joy of Element Collecting

I STARTED COLLECTING elements in 2002 and thought that in 30 years I might have most of them. Thanks in large part to eBay—and my own insanity—by 2009, I had assembled nearly 2,300 objects representing every element, the possession of which is not forbidden by the laws of physics or the laws of man. You have seen many of these precious items in this book.

To quote ABBA, "What a joy, what a life, what a chance!" OK, maybe being an international pop star is more exciting than the life of an element collector, but it has its moments.

I particularly enjoy finding oddball elements in unexpected places. Who would have thought that you can find very pure niobium (41) in very impure piercing shops—the kind that make you feel like disinfecting yourself after you leave. Or that Wal-Mart sells simple rectangular blocks of pure magnesium (12) metal, about as plain as can be, with no other function than to be used to demonstrate the fact that magnesium is a flammable metal. (They sell them in the camping section: You shave off a bit with a hunting knife, then use the attached flint to light the shavings and thus your campfire.)

Some elements can be experienced in large quantities, like the 135-pound iron (26) ball I keep in my office for people to trip over. Others are best enjoyed in responsible moderation—keep too much uranium (92) in the office, and people start asking questions. (Keep over 15 pounds, and the Feds start asking questions.)

Element collecting isn't a big hobby. Compared to the number of people who collect chemical compounds (minerals), polymers (Beanie Babies), or the same darn metals over and over again (coins), we element nuts are few and far between. Part of the reason is that even just storing your collection safely requires significant chemical knowledge. Sodium + damp basement = bang. But if you're willing to learn the ins and outs of each unique element, collecting them can be a tremendously rewarding experience.

See you all at periodictable.com, where I do my collecting and you can join the fun!

The author in his element(s). Shown on top of the world-famous wooden periodic table table are a small fraction of the author's nearly 2,300 samples of the elements and their applications.

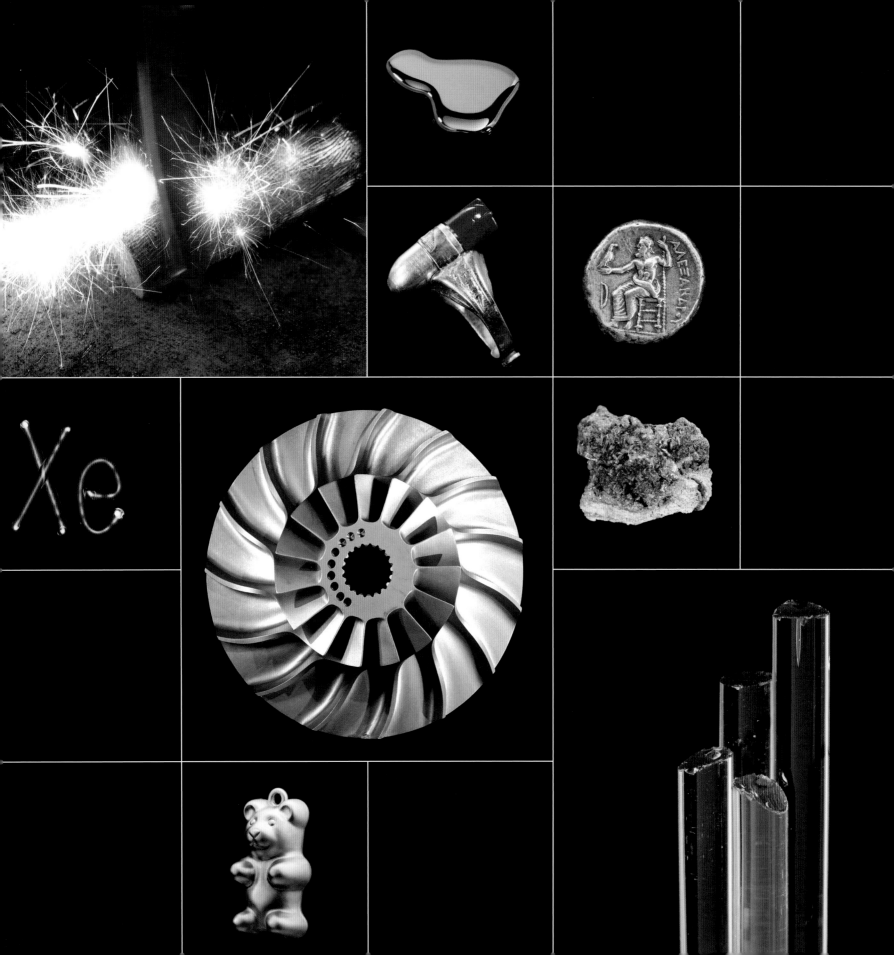